The Praxis of Product Design in Collaboration with Engineering

Wayne C. Chung

The Praxis of Product Design in Collaboration with Engineering

 Springer

Wayne C. Chung
School of Design
Carnegie Mellon University
Pittsburgh, PA, USA

ISBN 978-3-030-07046-5   ISBN 978-3-319-95501-8   (eBook)
https://doi.org/10.1007/978-3-319-95501-8

This Springer imprint is published by the registered company Springer Nature Switzerland AG
The registered company address is: Gewerbestrasse 11, 6330 Cham, Switzerland

Preface

This book is a culmination of more than 20 years of academic experience in teaching and advising industrial design and engineering students. I have had the fortune of holding three different tenure-track posts in public and private research universities. Each of the industrial design programs had similar curricular structures that were based on a design studio format. And at each post, the interdisciplinary collaboration between design and engineering was my personal and research interest. From this history, I have observed several key elements make significant differences in preparing young designers for an ever-changing world. Because of these experiences, I have the benefit of multiple points-of-view on how and why a particular blend of theory and practice is necessary for training future designers. The fundamental praxis of design, whether as an individual or in a team, requires a repertoire of actions and outputs so that better decisions can occur while advocating for the human in a built world.

This text's purpose is to share the unique perspective of product design education and interdisciplinary collaborations. These experiences are from a studio teaching perspective with case stories that highlight the chapter concepts in applied settings. These stories usually include industry collaborators or a funded research project. The subject matter ranges from one-of-a-kind research devices, future mobility systems, medical systems, and various other projects for a wide range of clients. The unique people and stories may bring a new understanding and appreciation of the challenges when attempting to combine beginning designers with different knowledge and skill sets. Having the fortune to have worked in the professional consulting industry prior to three different university institutions over my career, this text will provide a personal lens not typical of others. Over time, subtle shifts occur and only noticeable with longitudinal reflection and introspection. Hopefully, some of these experiences may be relevant to the reader's own journey, act as a guide when interacting with different disciplines, provide some insight into how designers work and think, or simply be reflective anecdotes because of your own history.

The design process and design thinking have shown the ability to address simple to complex problems in different ways than the scientific method. Complex and open-ended problems can be characterized as social, system, or intertwined situations that do not lend itself to earlier conventions or rote processes. This text describes how designers and engineers may address the range of complex challenges through a combination of flexible processes and activities. It is the culmination of "research of design" and "research by design" that will convey the importance of human-centered focus that grounds the decisions. But more importantly, the ability to work with people through constructive processes evokes and enables new ideas and better solutions.

This book recognizes both product design and engineering as disciplines that address complex products and systems in their prescribed ways. Much of the text is from an industrial design perspective. More so, it is about the practices, processes, and mindset necessary to achieve a common goal when working in problem spaces ill-defined that demand appropriate innovation. The breadth of experiences and stories is based on collaborative projects when working together towards a singular goal was a necessity. It is through this lens that other disciplines may glean new perspectives throughout their careers.

To overly simplify, both designers and engineers want the same thing. We want to provide a highly desirable product or system that can be intuitively used to perform the tasks necessary to achieve a desired goal. However, the differences in how to teach and achieve this are not self-evident at the higher education level. Curricular goals, learning outcomes, evaluation, reward systems, and the process to teach a designer are very different. And this is most pronounced in the early years of higher education, a critical and impressionable point in a young person's trajectory. It is this friction and tension that bears insights to the benefit of each discipline's pedagogy and practices.

To make some distinctions, the US Labor Department, Bureau of Labor, defines the job descriptions of engineers and industrial designers as very different occupations. In the United States, at the university level, each discipline is most likely housed in a different college. This text focuses primarily on the collaboration between industrial designers, mechanical engineers, and biomechanical engineers. Additionally areas of product engineering, product development, and engineering design may find this text relevant.

Industrial design sits in a unique space between educational theory and professional practice. In that regard, industrial design can be found within art, architecture, engineering, or even liberal art colleges within a university structure. These home positions were predetermined a long time ago at the design program's inception. The placement is likely because of legacy decisions made several decades ago dependent on the administration, funding, and space and resources.

Some definitions of design terms need to be stated explicitly because of the over-proliferation and noninstitutionalized nature of the design field. In the case of this book, the term industrial design is used to describe primarily a tangible, physical, three-dimensional object that exists in space. The traditional term industrial design is steeped in historical precedence dating to the 1930s in the United States' first degree program at Carnegie Tech (Carnegie Mellon University). However, the majority of the book will use the term product design. This is done purposefully to signify everything tangible and intangible from a graphic interface, a component of a computer system, and to physical objects. Modern-day programs have changed their names and degrees to reflect the growing need to educate and train designers who can do both physical and digital design. Thus the term product design can cover an incredible range of competencies. To illustrate the lack of oversight and standardization, graduates with industrial design degrees work for digital companies like Google, IBM, and Facebook and have their business title and cards as "Product Designer." In contrast, designers who have the same title but have never used a table saw or sanded bondo will also use the term product designer to describe themselves. Thus the confusion and irregularities abound and industrial designers invariably have to explain themselves as both 2D and 3D digital and physical product designers.

This text will use the term product design to describe the most recent transition. Today's product designers are no longer relegated to working in a model lab crafting high-density foam models. To be valuable and useful on a project team or any business entity, they are prepared to deliver both physical and digital assets. Many designers trained in classic three-dimensional model making, interaction design, ergonomics, etc. find themselves working in technology-, service-, or strategy-driven corporations. And these corporations may never produce a tangible artifact. But the term product designer is most apt because they are concerned with the meta-level picture and are still problem solvers. And more relevant to today's climate is the need to build digital products and/or physically tangible products together. It is nearly impossible to ignore a digital interaction on a display or on the product itself. When the term industrial design is used, it is in reference and reverence to the classic discipline where the work output is traditionally three dimensional.

Fortunately, the term engineering and the subcategories have strong consistency and shared comprehension. Most designers and engineers know the differences between mechanical engineering, electrical engineering, computer engineering, and industrial engineering. In the case of this book, the primary interactions are with mechanical, biomedical, and some electrical engineering. However, from working with various engineers over the past two decades, their desire to understand and learn why product designers are able to see things differently, approach problems from an obtuse direction while finding new insights to problems, has been a constant. There has been in-depth investigation on design, design thinking, and design engineering educational processes. Chapter 6 only touches upon a few significant engineering authors and their research. In no ways exhaustive, it represents a cross-section of engineering educators who recognize that the manner, methods, and content of teaching require more than the current standard. My prior engineering collaborators are continually trying to instill project-based learning earlier and earlier in the curriculum. This confluence towards introducing complex projects or open-ended challenges, what Rittel calls wicked problems, generates and graduates more agile thinking and resourceful problem solvers.

This extending arc of industry nomenclature is reflected in the relevance and complexity of today's design climate. As each product design and engineering program evolves, it is in response to an ever progressive need to educate for the future industry and societal demands. Terms like interface designer, user experience designer, and human-computer interaction designer all create an overlapping kalei-doscope of professions. Even the design process and methods may be very similar. The most significant distinction is not necessarily in the processes, but the activities and subsequent output and outcomes. And one of the purposes of this book will bring some clarity and open new conversations to how designers and engineers can benefit from understanding each other's educational experiences and pedagogy.

The field of design used to be relegated and perceived as a tactical piece of a larger corporate brand or organization. However, it is now commonly applied for strategic and system-level impact because of the studio taught philosophies, prin-ciples, and processes. Contemporary designers are now looked upon as agents for transformation in organizations and communities. And thus they are the perfect catalyst in technical industries but also in the humanities. A studio-trained design-er's mindset uses the human-centered pedagogy while converting abstract concepts to tangible mediums. This generative design and thinking process is at the heart of why and how design has taken center stage in keeping the artificial world, human.

Pittsburgh, PA, USA   Wayne C. Chung

Acknowledgments

I am so thankful for my immediate family, Brandi, Julia, and Weston. They were so supportive of my labors to compile something relatively coherent.

I have to thank my father, William K. Chung, Ph.D., who lived and acted by principle. Retrospectively, I'm glad he lectured my brothers and me so often. Eternal thanks to my mother, Theresa, for being the rock I needed while still being a mom. Thank you to each of my three brothers Willis, Wilbur, and Wally. Whenever I needed medical, academic, legal, or life advice, it's nice to know you're covered.

A sincere and continued appreciation to all of my colleagues at OSU, GT, and CMU. This book would not be possible without free dialogue, critical debates, and all the incredible talent around me. Special thanks to the Chairs of my home department who entrusted me with the next generation of designers and innovators. Thank you Jim, Lorraine, Dan, and Terry. And of course, I want to thank my students who I've had the pleasure teaching and getting to know throughout these years. Your energy and talent bode brightly for our future.

Contents

Chapter 1
Vision and Tensions That Elevate Product Design Development

Design as Collaborative Activities and Output

Product design is at its best when working with other disciplines. A studio-trained designer brings a unique mindset, skill sets, and a breadth and depth of knowledge that is technical, experiential, and human-centered. Because of the manner in which they were educated, product designers are at the nexus of both practicality and possibilities. It is more than their knowledge of material and manufacturing. Rather it is the combination of studio environment experiences, methods of how to learn from the world around them, proficiency to produce a range of tangible artifacts in various mediums, and ability to dynamically synthesize actions and output for better decision-making and innovation. Also, the type of education and the manner in which they are taught enable them to be generative thinkers and learners.

When embedded with engineering or product development teams, a studio-trained designer provides the unique composition that creates the right amount of tension to shift the norm. The tension occurs when the product designer asks why molding a clear polymer with particular tolerances has not been done before in the desktop computer industry. New categories and standard of products emerge when a design team demands aluminum and glass to be integrated into everyday devices. The qualitative nature of how product designers approach problems can also generate possibilities and impossibilities. But these can produce positive effects by pushing the boundaries of what has not been done or considered before. And it is the inclusion of many people and stakeholders voices into the design process that can create internal team tension, but eventually innovation. How a product designer is educated and subsequently how they interact with the world inevitably creates healthy tensions between many of the team disciplines. And it is this catalyst within a team that a product designer finds themselves as a proponent for the human end user, an evangelist for new and untried ideas, and a leader who believes so devoutly in the process and methods that they find themselves propelled into critical roles throughout the corporate and institutional levels.

© Springer Nature Switzerland AG 2019
W. C. Chung, *The Praxis of Product Design in Collaboration with Engineering*,
https://doi.org/10.1007/978-3-319-95501-8_1

Studio-trained designers and engineers find themselves submitting new and uncomfortable possibilities. Otherwise, there would be no innovative visions of and for the future. A future state where people and product not only complete a task or goal but an appropriately designed system state where all the actors: persons, product, place, and context are considered would become feasible. In addition to balancing all of these interrelated elements, product designers are also naturally creative. And this is not limited to problem-solving. They attempt to push the envelope of possibilities before a suitable resolution can be agreed upon. The process of product design inherently stretches boundaries, challenges the limits, and circumvents the norm. How can we achieve this healthy tension internally and externally? And how can we teach this method and mode of thinking?

Product design is a combination of purposeful activities and outputs. The combination of these two complex elements converges and diverges to frame a problem and a multitude of solutions. It because of the manner in which the design process is executed, that studio-trained designers are apt to think, behave, and create in highly unique ways compared to their technical partners. In a simplified fashion, the design process steps would be researching that problem, creating a range of concepts, and using iterative feedback loops to arrive at a solution. This oversimplified definition can easily describe other disciplines like engineering or even music composition. However, the distinctions and differences are found in the fundamental approaches to teaching and the resulting professional practice. The accepted professional definition is stated by the Industrial Design Society of America. According to IDSA, "Industrial Design (ID) is the professional service of creating products and systems that optimize function, value, and appearance for the mutual benefit of user and manufacturer." If we replace the industrial design term with mechanical engineering, this definition would still hold true. Even as we read the definition's subsequent lines, an engineer could still fit the same criteria. "Industrial designers develop products and systems through collection analysis and synthesis of data guided by the special requirements of their client and manufacturer. They prepare clear and concise recommendations through drawings, models and descriptions. Industrial designers improve as well as create, and they often work within multi-disciplinary groups that include management, marketing, engineering and manufacturing specialists."

This is the constant conundrum of the product design discipline. It is a field that straddles arts and science, aesthetics and ergonomics, business and intuition. But the power of product design is greatest when working with others and for others. Thus, the altruistic nature of the field and the end product culminates in blended anonymity. So why does this modern day definition still not show the distinct differences in each field?

Looking back at pivotal moments in history may allow us to capture the distinction of when the field was born. The Industrial Revolution and the manufacture of consumer goods are considered the initial time when the practice and term, industrial design was formed. Heskett, J. (1985) With the modernization of mechanisms, machinery, and relatively consistent power sources, the ability to scale and deliver standardized, mass-produced goods was possible. As the population and society required these products and goods, so did the demand. This march of mechanization shifted the paradigm from how traditional craftsmen would create a one-of-a-kind object to mass manufacture of many units. The mid-eighteenth century required new ways of thinking and training to deliver this multitude of goods. Great Britain is recognized as the birthplace of the Industrial Revolution. But Germany plays a significant role in the education and history of today's industrial design system. Established in 1919, the Bauhaus was a German design school infamous for the residents in practice and philosophical approaches. Founded by Walter Groupius, the school was dedicated to all arts, including visual arts, architecture, design, and design education. Its influence is found in graphic design, interior design, typography, industrial design, and essentially what was called, applied arts of the time. Their first-year or preliminary courses taught basic principles of design, such as color theory, form theory, and life drawing. One of the overarching principles of the school was to unify art, craft, and technology. This serious and pragmatic approach seems to strip decoration and unnecessary ornamentation from buildings and objects. Companies such as Braun and Olivetti employed many graduates of the Bauhaus. And their visual design and overall philosophies are self-evident in contemporary designs. Apple Computer's designers and consulting agencies such as frog design show the enduring spirit of this Bauhaus philosophy. Besides the course curriculum, the place and space of learning were very significant. What they called workshops, today we call design studios. This is where the education and training was combined. The process and work were all in one space—as were the designers themselves. Subtly different from an apprentice and master setting, this was a hybrid of an art studio and practical training environment. Students were able to experiment, self-reflect, and make adjustments to the feedback and input from instructor and peers alike. Different from a science lab, today's studio spaces are modeled similarly where students have access 24/7 and make their "classroom" their own living laboratory. Practicing their craft, execution, and reworking the work defines a designer as a practitioner. This aspect is covered in more detail in Chap. 2.

Fast forward to modern day, US industrial design history. Carnegie Mellon University's School of Design is credited for establishing the first program in 1934 by Robert Lepper. Lepper was a graduate of Carnegie Tech in 1927 and worked in Europe exploring contemporary art. Much of the Bauhaus legacy is still is apparent in twenty-first-century industrial design education and professional studio practices throughout the world.

Present-day product design education and practice are still concerned with creating products that optimize function, value, and appearance for the benefit of all stakeholders involved. Stakeholders include both the user and the company producing goods or a system. One of the critical but different elements from other fields is product design's responsibility to the human and their overall experience. This experience may be with a tangible object, a digital display, or an environment. Therefore, the primary emphasis and point of entry to projects and problems are human-centered. Whereas a business person is interested in reducing costs and maximizing profits, the product designer is concerned with the aesthetic and brand language. The engineers producing the widget are concerned with manufacturing tolerances, assembly, and optimization, while the product designer is also concerned

Fig. 1.1 Tim Brown's
three spaces of innovation.
Image courtesy of IDEO

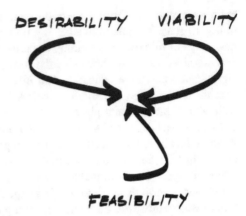

with ergonomics, user-to-machine interactions, and the affordances the widget is expressing in various system states. Therefore, the designer's research processes, general inquiry, generation of concepts, testing, and subsequent decisions always focus on understanding the human experience in relation to product, place, or other persons. Figure 1.1 from the book Change by Design (2009) is a representation of the overlapping factors necessary to enable success. Brown describes these factors as follows:

Feasibility: what is functionally possible within the foreseeable future
Viability: what is likely to become part of a sustainable business model
Desirability: what makes sense to people and for people

A subsequent IDEO Venn diagram equates Feasibility to Technology, Viability to Business, and Desirability to People. This noun-based analogy of People, Business, and Technology is a guide to where emphasis and energy can be focused. Between each of those overlapping areas exist differing types of innovation such as process innovation (Business and Technology), functional innovation (People and Technology), and emotional innovation (People and Business). Design as a discipline enters this space through understanding people first, rather than an emphasis on the other domains of technology or business.

Product designers are concerned with the human and user as their primary constituent. Even though they are trained in materials and processes of mass manufacture goods, technically adept at 3D computer-aided drawing applications, and understand business principles, they enter the problem with the person in mind. Combined with education and experiences in a range of visualization and fabrication techniques, a product designer is technically and cognitively prepared for a wide variety of challenges. More specifically, a designer is able to visualize research, concepts, ideas, abstract systems, and concrete objects at varying fidelities. And it is this ability for agile thinking utilizes processes and methods to understand the overall situation. It is the culmination of these skills, activities, and mindset that other fields see this as a method to achieve breakthrough innovations. These activities coupled with the unique workplace and dynamic have been adopted by other disciplines and this has been popularly coined "design thinking."

This type of work process has garnered both great attention and scrutiny from other fields. Disciplines such as engineering, business, medicine, and military have employed more designers. And this is regardless of that industry's end output. Designers work in product, commodities, digital, services, and systems industries. Regardless of the tangible or intangible output, designers are being placed in all types of roles. In 2016, IBM dedicated their resources to hire an army of more than 1250 designers within a span of a few years (Cao, 2016). They invested more than $100 million towards design and understanding user experience across their business (Wilson, 2014). Financial corporations such as CapitalOne acquires design consulting firm, Adaptive Path to leverage the assets and culture of design. And the more traditional product design corporations such as Lunar are acquired as smart investments by the likes of McKinsey & Company (Ha, 2014). The article by Kyle Vanhemert (2015), "Consulting Giant McKinsey Buys Itself a Top Design Firm," mentions the critical and pivotal importance of design. It references the incredible story of Apple's underdog and floundering stature. And credits design as one of the most powerful examples of turning around their fortune. So what is so special about the design process? Why is everyone looking to design as the panacea for everything from flat profits to addressing wicked problems? As expanded upon in the later chapters, the activities and output during the process of design allow for a synthesis of team members. But explaining the details and nuances of how product designers are trained and how they work may bring light to their unique stature.

To better understand product designers, we need to look at how they are trained and educated in the studio environment. How is "design thinking" inculcated in students? The design process allows for both exploration and focus through a longitudinal time span. It is within this process that designers can find stories that are the basis for insights and inspiration.

Design Processes in Design Education

To better understand the process and mindset of a designer, the traditional design process needs to be explained. Most design process descriptions have at least four distinct stages. The double diamond process diagram developed by the Design Council UK (2018) is widely used to describe these activities (Fig. 1.2).

This interpretation is useful up to a point. Most open-ended and long-term project-based work does not follow this oversimplified process. Expounding on this diagram further is Dan Nessler's detailed version of the double diamond configuration (Fig. 1.3). Upon inspection, there are key phrases that best sum up critical intentions of design. The first to note is the top two phrases: Designing the right thing, and Designing things right. Though subtle in wording, these are two distinctly different phases. And each phase requires a different set of skills, different

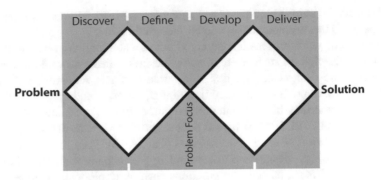

Fig. 1.2 Double diamond design process

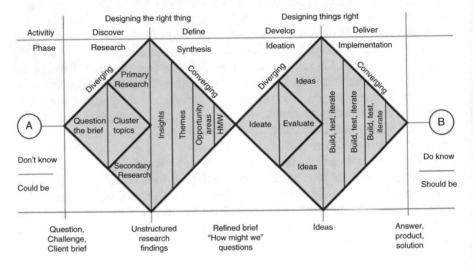

Fig. 1.3 Dan Nessler double diamond detailed design process

training, and different mindset. This is one of the most important aspects that we teach freshman, new graduate students, and workshop participants from corporations not familiar with the design field. To actually start making something in the model shop or on a computer application, a designer needs to have the bigger picture in mind before manipulating matter. They need to be designing the right thing before they apply aesthetics, ergonomics, or manufacturing principles to the product.

As shorthand, this left diamond is considered meta, high-level design thinking or Design with an uppercase "D." Even if a client presents the problem, designers will question the brief and zoom out or back out to understand the full context of the problem. This is called framing and reframing the problem. Designing things right would be considered a lowercase "d" or the micro level of details that include form, material choice, and aesthetics. In the corresponding right diamond, you will see the areas of ideate, build, test, iterate sections. The traditional field of product design would be recognized for this process and application of craft and skills. Methods and techniques for this interplay of problem-solving, systems-level thinking, identifying stakeholders, and considering societal and cultural aspects are more fully covered in Chap. 3. This is being recognized as one of the more powerful aspects of design and the design process. The praxis of this combined "Design" thinking (meta-level design processes and thinking) coupled with traditional "design" (tangible form giving, aesthetics, material selection, ergonomics) is expounded upon through the book with case studies and is labeled as "D/design" when recognizing the intertwined activities and outputs.

So what may be the differences in relation to an engineering design process? Taking NASA's engineering process as an example, the steps are listed as follows:

1. Identify the problem
2. Determine criteria and constraints
3. Brainstorm solutions
4. Generate ideas
5. Explore possibilities
6. Select approach
7. Build prototype
8. Refine Design (NASA Education, 2017)

In comparison, the left diamond in the design process seems to be already assumed when comparing to the engineering process. Thus one of the significant differences in the design approach is the initial and constant questioning the given problem. One may argue that using a time continuum, the previous step of questioning the original problem has already been completed. The contrast to this argument is that designers are constantly diverging and converging in their thinking and processes to test the boundaries up until they commit to a single solution. This type of mindset for diverging ideas sets may designers apart from an engineering mindset.

The purpose of engineering is to arrive at a plausible, manufacturable, and effective solution. This solution has to perform under various real-world criteria such as stress and strain demands. As Ullman rightly states in one of his chapter book headers: "As Knowledge is Gained, Design Freedom is Lost" (Fig. 1.4).

Comparing the engineering and design steps could still be argued as quite similar. Is it mostly semantics and shifting of some of the stages or steps? The evidence that engineers and designers are different from one another cannot be easily seen from comparing a process diagram. The keyword may be time. Some of the main differences in what each discipline perceive as success could be how designers and engineers are taught and think differently. Engineering grapples with business and technology to arrive at an optimal solution. And if this can be done within a shorter amount of time, then the business side of the PBT Venn diagram achieves an added benefit.

Both disciplines perceive the term success in different ways. If Design is concerned with people as its primary constituent, then ergonomics, aesthetics, and the

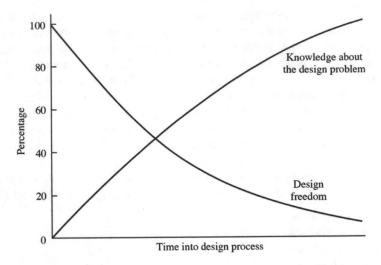

Fig. 1.4 Ullman design paradox

user experience are paramount. If Engineering's primary concern is technology, the ability to produce a system that accomplishes its goal with a high success rate is paramount. Even though both disciplines use comparative and competitive analysis, both may interpret and imagine things quite differently because of their role and responsibilities.

Innovation Through Process and Project Oriented Teaching

Many title texts in engineering education are dedicated to innovation. A search through Wiley Online Library of Journal of Engineering Education from 2000 to 2018 on the term "innovation" in the title of the publication field returned over 2100 journals and 1100 book results. The 2000–2001 period showed only 154 Journals and 15 Books. A plethora of information, knowledge, and wisdom abounds in these texts. They provide techniques, methods, and examples for building teams and recommending processes that lead to innovative solutions. However, innovation, and the ability to take risks are both arguably a combination of nurtured (learned) characteristics and nature (inherent personality). This is where one of the greatest distinctions between the fields may exist. Design students are rewarded for imagining what could be, unique approaches, untried solutions, and essentially being different. This is a key aspect of the designer's process as noted in Fig. 1.3. Engineering students are rewarded for the correct solution and optimal designs in relation to business, technology (feasible materials and manufacturing) early in their studies. They can achieve these goals with greater probability by reducing unknowns and constrained product criteria. This type of reward system is inculcated through their educational curriculum. An investigation into several mechanical engineering curriculums show that the first courses are Calculus, Thermodynamics, Physics, Electromechanics, Fluid Mechanics, etc. A studio-trained design student's early, first 2-year courses typically are freehand drawing, design studio, art or design history, and photography. Depending on the school, the only subjects that may be similar between both disciplines is a humanities course such as writing, history, or psychology. The only course that may be the same in content is a computer-aided drawing class. However, if you compare the primary courses the major emphasis is on the following:

Design: Design Studio, Visualizing, Design Studies
Engineering: Calculus I, Calculus II, Physics I

There is a correlation between the type of reward system that sets the stage for an impressionable student. Early engineering course problem sets are typically math based on one correct or optimal answer. Design emphasizes understanding people, sociology, and humanities. Their design courses require and reward based on a broad discovery of their own style in relation to others.

And mentioned earlier, the differences in how the disciplines portray their process are quite telling. In both double-diamond diagrams, the right side captures the activities and outputs related to Develop and Deliver. Engineering has the tendency to always start on the right side, Designing Things Right framework. How to teach problem-solving is obviously important. However, there are many times when the big picture is lost and making a better widget, is not better. As alluded to in the IDEO diagram, if Design is a proponent for People, what is Engineering's calling? By the earlier descriptions of the first-year students' educational coursework, it would seem Technology and Business prevail.

However, the product design process of using both double diamonds reward students for investigating human behavior, perception, and values. This philosophy and early educational imprinting lead to a sensitivity to setting the problem in direct context with users, stakeholders, humans, and society. This type of contextual inquiry and generative thinking aligns with the highly unique nature of every problem. Stanford d.school has aptly used the keyword, "empathy" in their design process. This rich word represents the nature of design thinking as an endeavor to understand people. If one has empathy, it implies essential qualities of placing yourself in other people's mental and physical situations. If we review the educational requirements and reward system of engineers, there is little emphasis on understanding humans in the early parts of an engineering curriculum. This is very evident in the early years. With the exceptions of cornerstone or capstone projects, there is a high possibility an engineer may not be given an opportunity that places the human condition or needs as a central factor to a project's success. And experienced designers and engineers can attest that the human variable is the most difficult to predict and build for. This very fact begs for some introspection on what, when, and how to teach this human aspect in a technical field. Donald Schöen's seminal book, "The Reflective Practitioner" was an introspection on how to address and balance technical knowledge with a more reflective practice. Schöen's call for a new epistemology of practice states the ever-present challenge of educating students as a "crisis of confidence in professional knowledge" (Schon, 1990).

In the varied topology of professional practice, there is a high, hard ground overlooking a swamp. On the high ground, manageable problems lend themselves to solution through the application of research-based theory and technique. In the swampy lowland, messy, confusing problems defy technical solutions. The irony of this situation is that the problems of the high ground tend to be of relatively unimportant to individuals or society at large, though their technical interest may be, while in the swamp lie the problems of the greatest human concern.

As the world develops more rapidly and adopts technology at every turn, the champions for useful and enjoyable interactions will require a combination of skills and sensitivities. Products, systems, and services that succeed are ones that deliver beyond the basics, and build a relationship of trust, confidence, and cohesive experience. It is this concern for the human condition and emotions that drive the proliferation of design thinking processes. And as you read through this book, the central argument is for continued and concerted investigation of people in situ so that appropriate advances and innovations may occur. Human behavior can be at odds with machine interfaces, the internal workings of machine learning, and artificial intelligence output. As technology accelerates at such a rapid rate, so does the logic and interactions of the artificial. Thus, it is of critical importance that we are attuned to the human condition and how human agency is being affected.

The following case study is a collaborative team project with engineers and a designer developing a better method and delivery system for drinkable water in remote and disaster relief situations. This entrepreneurial story starts with a highly defined goals but through heavily iterative and process driven activities, the team eventually builds highly viable and marketable solutions. Subsequent book chapter case studies show the different potential in using the design process and their activities and output as building blocks for collaboration and innovation.

Case Study: Rorus Inc.

Four engineers and one industrial designer formed a biomedical team for their capstone project. One particular civil and environmental engineer, Corrine Clinch had a vision: to provide drinkable water in third-world countries, disaster relief zones, and ultimately, at home. Clinch had recently returned to complete her master's degree after public health work in Ghana, South Africa, and Haiti. She experienced first-hand the real limitations of various water purification products and systems. Existing filtration systems perform well in the lab or when used by a well-informed user. However, education of nonvisible pathogens and harmful chemicals are difficult concepts to comprehend for the intended users. Even when boiling water or using certain filters still left water cloudy or with similar turbidity post-process. Uninformed people did not understand the benefit and safety they were achieving by doing these extra time- and resource-demanding processes.

The multidisciplinary team of mechanical, chemical, material, civil and environmental, and industrial design students was assembled for this highly complex challenge. Device for Emergency Water or the DEW team's two primarily challenges was the technical aspect of fast purification of any water source; and secondly the human challenge to change behavior through the product use. Education and cultural change is the ultimate goal for safer drinking behavior. But the DEW team needed to compete against UV disinfection, chlorine tablets, ceramic filters, and microfiltration filters. And the team had access to a silver bullet. Clinch had been working with a researcher using nanotechnology material that had silver ions baked into the filter. Rather than forcing contaminants through smaller and smaller holes, the new nanomaterial worked like a magnet attracting and attaching to the filter lattice. And in the case of the newly developed nanomaterial, the holes were a factor of 100 larger than competitor filters. This translated into fast decontamination and very usable and high flow rate. The team knew they had a competitive advantage on the technical side. The challenge was how to package and deliver such a new product system. The team identified the highly definitive problem and clinical needs:

- Consumption of unclean water is a direct threat to public health in areas with poor infrastructure.
- Consumption of unclean water can cause dehydration, fever, and muscle aches, and death.
- Approximately 3.4 million people die each year from a water related disease, and 780 million people worldwide lack access to clean water (WHO, 2012).
- We need a solution that provides safe, accessible drinking water for disaster relief and low-resource areas
- We intend to provide short-term clean water at a lower cost than current industrial products and offer a large-scale deployable solution.

However, even with this new material technology, the DEW team's challenge had only begun. Over the course of the two semester-long project, over 50 physical models were built and tested for both technical performance and human usage in various environmental conditions. How the water is introduced to the filter container? How much water could it hold in relation to the filtering output? Does a person filter at the water source or do they bring that water home to process? How can women and children haul several kilograms of water in underdeveloped environments over long distances? These very human situations posed significant design interaction and decision challenges. After 9 months of extensive usability studies, the team incorporated the first viable solution (Fig. 1.5), that would serve as the entry point for a startup company that would raise more than $1.1 million in funding.

Original Device for Emergency Water team: Anne Alcasid, Corinne Clinch, Uriel Eisen, Elizabeth Ha, Yixing Shi, and Christina Yoon. Advisors Dr. Conrad Zapanta and Dr. Theresa Dankovich.

Ilana Diamond of Project Olympus, a startup incubator program, attended the final presentation of the BME projects. And Diamond encouraged Clinch to pursue this unique solution to a global problem by entering the incubator program. This program provided micro-grants, work space, and connections to regional and

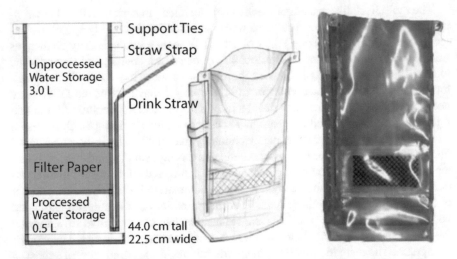

Fig. 1.5 Proposed DEW final design diagram, concept drawing, and final physical prototype

national business contacts. The company renamed as Rorus, Inc. with Uriel Eisen, industrial designer and Clinch as the founders. Their original goal to serving beyond the disaster and temporary cases expanded to home and family use. One of the prevailing reasons was that the nanotechnology was so effective, that it could safely filter water for a family of six to an entire year. Very quickly they realized competing with current companies, regardless of their technology and scaling for manufacture economies, they would need even more competitive advantages. This came in the form of good design through living through the design process. Clinch and Eisen continued their continuous prototyping and taking these into the wild. The following images and Fig. 1.6 represent the field research in situ in rural parts of Nigeria and India and their current design solutions. Clinch's description of an OpenIDEO challenge best characterizes the thoughtful design that deals with the technical as well as the human needs.

Product Solution: At Rorus, our new filtration nanotechnology enables new design paradigms to prevent misuse and failure in the field. The Rorus cartridge provides immediate and safe drinking water from any fresh (non-salt) water source for a family of six for a full year (>7500 L). Most importantly, it can work without the education and daily management required by solar, chemical, microfiltration, and other purification options.

Our human-centered design process has led to two product offerings, a filter backpack and an adaptable cartridge that works in any locally available bucket. The backpack option is best for disaster situations during the Indian monsoon season and for families who regularly carry water from a remote source. The bucket option is best for families who regularly have piped water in or near their home. The filter and spout come as a unit that takes <3 min to be installed in any plastic bucket or container.

Fig. 1.6 Rorus field research and product testing

Product Benefits:
Rorus focused on solving the specific issues where other water purifiers fail.

- Instantaneous purification.
- Fast flow delivers immediate clean water which ensures continued use, not straining the user's patience with waiting periods. Because there is water on demand, there is no need for storing water in buckets or containers where it often gets recontaminated.
- No maintenance requirements.
- Simple design reduces opportunities for failure due to human error. Our filter requires no energy input or maintenance steps such as backwashing, cleaning, measuring dangerous chemicals, or timing operating steps. Users instantly get clean water by filling the container and opening the tap.
- Automatic shutoff.
- When the Rorus Filter is past its useful life, it automatically clogs, preventing overuse.
- No contaminants missed.
- The Rorus Filter removes bacteria, viruses, and protozoa in compliance with WHO standards. Users stop trusting microfilters that do not work on viruses when they still get viral waterborne infections and stop trusting chlorine tablets when they still get cryptosporidium. Complete and trustworthy purification is essential to both keeping users safe and ensuring continued filter use.
- Refreshing taste.
- The Rorus Filter also reduces turbidity, taste, odor, and chemicals. This means that users enjoy the taste and trust their water while not adding any taste as chlorine tablets do. Once again, this encourages adherence and promotes safety.
- Easy to deploy.
- Each Rorus Filter weighs <4 oz. and is approximately the size of a 1 L water bottle. With a durable physical structure and all sterile surfaces physically protected, it is easy to transport by plane, truck, or even in your backpack. One person can easily carry 200 years of safe water—equivalent to the cargo capacity of Boeing 747.

Fig. 1.7 Rorus product
line of filter core, Rorus
spring, and filterpack.
Clinch & Eisen (2018)

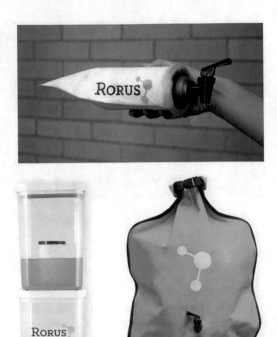

– Affordable cost.
– Each liter of water costs less than half a rupee (0.4 rupees per liter). This is
  cheaper and more sustainable than packaged water in India and most other
  filters.

Technology: Our technology works differently than other physical filters. Rather
than working like a net or a sieve to catch pathogens, our media works like a magnet
to attract and retain pathogens of any size. Independent testing has verified that this
removes 99.99% of protozoa and bacteria as well as 99.9% of viruses without any
leaching of silver, chlorine, or halogens. By a unique combination of technologies,
it also reduces 95% of arsenic and lead contamination and automatically prevents
biofouling, or any growth on the filter's surface (Clinch, 2016) (Fig. 1.7).

Specific education and behavior change is the ultimate goal for healthier living
in remote or rural areas. And even with a great product and viable technology, this
can happen depending on other factors. And with any startup company attempting
to shift the paradigm, constant assurance and a form of calculus are necessary to
enable the three areas to coincide: people, technology, and business.

References

Brown, T. (2009). *Change by design: How design thinking transforms organizations and inspires innovation*. New York, NY: Harper Business.

Cao, J. (2016). *Bloomberg News*. Retrieved February, 2017, from https://www.bloomberg.com/news/articles/2016-07-25/ibm-hired-hundreds-of-designers-to-figure-out-what-customers-want.

Clinch, C. (2016). *OpenIDEO Water and Sanitation Challenge*. Retrieved February, 2018, from https://challenges.openideo.com/challenge/water-and-sanitation/ideas/the-rorus-filter-pack-simple-clean-water.

Clinch, C., & Eisen, U. (2018). Retrieved February, 2018, from https://www.rorusinc.com/.

Design Council UK. (2018). Retrieved February, 2018, from http://www.designcouncil.org.uk/news-opinion/design-process-what-double-diamond.

Ha, A. (2014). Retrieved February, 2018, from https://techcrunch.com/2014/10/02/adaptive-path-acquired-by-capital-one/.

Heskett, J. (1985). Industrial Design (World of Art), Thames & Hudson.

NASA Education. (2017). Retrieved April, 2018, from https://www.nasa.gov/audience/foreducators/best/edp.html.

Schon, D. A. (1990). *Educating the reflective practitioner: Toward a new design for teaching and learning in the professions* (1st ed.). San Francisco, CA: Jossey-Bass.

Vanhemert, K. (2015). *Wired Magazine*. Retrieved February, 2018, from https://www.wired.com/2015/05/consulting-giant-mckinsey-bought-top-design-firm/.

Wilson, M. (2014). *Co.Design Magazine*. Retrieved February, 2018, from https://www.fastcodesign.com/3028271/ibm-invests-100-million-to-expand-design-business.

World Health Organization and UNICEF. (2012). *Progress on Drinking Water and Sanitation: 2012 Update*. New York, NY: WHO/UNICEF Joint Monitoring Programme for Water Supply and Sanitation.

Chapter 2
Mindset and Modes of Design Studio Education

Understanding People Before Users: Human-Centered Design

Chapter 1 dedicated much of the text to drawing out unique aspects of design and engineering. However, both disciplines are ultimately chasing the same thing: good design. Both disciplines enter the field intent on delivering products and systems that satisfy users who have certain tasks and goals. And if they are completing these tasks and goals with efficiency and pleasure, then the product solution is that much more successful. The differences may be in the manner each discipline is taught and subsequent philosophies to achieve their desired problem solutions and user state. The subheading of this chapter heading explicitly uses the term, "people" before committing to the right side of the double diamond or the "user." Similar to the design process, product design attempts to change the perspective to a higher conceptual point of view before a narrow focus. Though there are subtle differences between the terms, people, and user, it inherently changes the intent of the designer's approach and opportunities. This initial mindset and divergent thinking allow designers to frame the problem rather than prematurely focus on the obvious low-lying solutions.

Most studio projects are considered journeys. There is not a singular, correct answer. Rather a range of contextually developed solutions as a result of applying a human-centered design process. And the raw materials and inspirations that students base their decisions upon come from the research participants. Other external research sources are used such as literature searches and competitive analysis of prior products and projects. However, the need to understand the primary constituents of the product they are trying to serve is paramount. Therefore, research in many forms becomes the critical component of a human-centered design process. Figure 2.1 shows the difference between approaching a project or problem through a person-centric aspect, rather than a user. When developing a research protocol for a participant who is defined as a user, then we assume there is a preconceived task and immediate problem or objective to be completed. This protocol may be perfectly

© Springer Nature Switzerland AG 2019
W. C. Chung, *The Praxis of Product Design in Collaboration with Engineering*,
https://doi.org/10.1007/978-3-319-95501-8_2

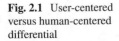

Fig. 2.1 User-centered versus human-centered differential

Users	People
Task	Goal
Place	Context
Interaction	Experience

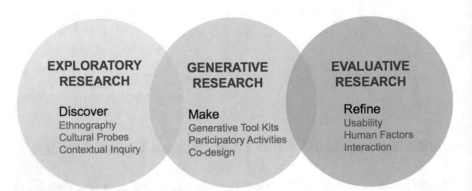

Fig. 2.2 Design research methods model, Hanington

valid toward the end of a project where critical ergonomics, efficiency, and optimization may be the end objective. But for the initial start of a project, designers need to be respecting people, as people. This research process is purposely broad in its approach. The importance and essence to this contextually driven method is to draw out personal stories and experiences in relation to the problem space. And these qualitative experiences accept the participant as the expert. It is the phenomenological approach that can make a difference in designing the right thing.

Hanington simplifies the main research areas that designers usually undertake as: exploratory, generative, and evaluative (Fig. 2.2, Hanington (2007)). Exploratory research is a designer's first entry into the problem space. The purpose of this phase is to understand the person in context/environment where the product is being used for a given task. It is from this ethnographic inquiry through other people's points-of-view that designers gain a relatively comprehensive understanding and purview of the problem space. Designers will use a variety of methods such as observation, competitive product analysis, contextual interviews, trace measures, and other traditional behavioral research methods. These traditional behavioral research methods set up the following generative research phase. Referring to back to the left side of the double diamond design process, divergent thinking and methods are being used to garner understanding and material for potential opportunities. These opportunities are problem redefinitions or what one may call reframing and a basic understanding of the people, product, place, and interactions relevant to that problem space. A more extensive discussion and case study example of this mindset and framework are expanded upon in Chap. 4.

The exploratory phase is essential for designers entering a field or subject matter that is new or foreign. A prime example of the exploratory phase's research method potential is highlighted in everyday products and interactions. Crown Lift trucks is an award-winning company in New Bremen, Ohio. Design is at the heart of their products of powered industrial forklift trucks. Crown started design improvements on their pallet trucks by improved ergonomics and man–machine interfaces through user-centered research methods. Their history of user and task research dictating best interaction practices are renowned. The company is a recipient of multiple awards including IDSA's International Design Excellence Award. Referring to the lowercase d: design aspects, Crown delivers on appropriately designed operator seating positions, control interfaces, and tactile material choices. In addition, one of their important innovations was the side seated driving and operating 5000 series ESR Series. Thought to be counter-intuitive, this side-seated position lift truck exists because of observational research and other corroborating factors showing operators repeatedly leaning their entire body to the side, in order to view the pallet and load. Multiply this over several hours and repeated frequency, and it is clear that a significant paradigm shift was needed. The lift truck industry perceived this inno-vation as a game changer and considered this a significant design shift both in at the meta-level and human interaction level. It took the mindset of entering an environ-ment with unbiased eyes and suspending self-judgment to recognize that a better method for forklift operators was possible.

The Exploratory phase can be characterized by a combination of third-person research such as literature search and traditional behavioral research methods. The Evaluative phase can be characterized as testing, confirmation, and verification of design concepts. However, designers use the generative research phase as a form of discovery rather than proof. Generative design research is a method for eliciting participant experiences through personal stories, preferences, needs and unrealized aspirations. This method of research is not without controversy. Generative research, also called action research or participatory research, has had its detractors. Different from the scientific method whose definition states:

"The principles and empirical processes of discovery and demonstration consid-ered characteristic of or necessary for scientific investigation, generally involving the observation of phenomena, the formulation of a hypothesis concerning the phe-nomena, experimentation to test the hypothesis, and development of a conclusion that confirms, rejects, or modifies the hypothesis" (The American Heritage Medical Dictionary, 2007). Also, "… scientific research is performing a methodical study in order to prove a hypothesis or answer a specific question. Finding a definitive answer is the central goal of any experimental process." (Explorable, 2018). The scientific process is taught and ingrained early on in K-12 students and continues through to higher education. And in the professional realm, the scientific process is a necessity when faced with rigorous, repeatable, and reliable research. The efficacy of this process is not in question. However, the introduction to other methods of inquiry is necessary when a clear hypothesis is not available or formulated.

The Relevance of Contextual Understanding to Innovation

An interesting trajectory of change has occurred in the design research field over the years. User-centered research was born out of practical needs in the military fitting of clothing, gear, pilot aircraft seating, and to driving positions. This mode of research resulted in definitive measurements that were the basis for anthropometrics and ergonomics. And design was in service of this methodology and resulting guidelines. This expert-driven research was a necessity in an age of man–machine interface and efficiencies to determine practical design criteria and dimensions. Fast forward to today's design research, and the generative phase of co-creating is an essential component to engaging, understanding, and enabling subjects to participate in the design process. This is especially necessary and true because of the rapid proliferation of technology and other factors changing and affecting how people live, work, and think in their own habitat. A deeper understanding of the culture, society, and nuances of the market is necessary. Generative ethnographic research and new methods are being developed to be part of the co-creation process. By conducting generative research to understand a cultural subset, allows the people to make their own preferred future scenarios more plausible and possible.

Daedalus Design, a product and engineering firm and ThoughtForm, Inc. a communication and strategy design consulting firm copublished a research toolbox diagram. This diagram is very useful to describe what type of method might best suit the research in relation to the participants (Fig. 2.3). By using the semantic differential of X-Axis as: Ask the User versus Observe the User, and the Y-Axis as: User can articulate their needs versus User not yet conscious of needs. This 2 × 2 matrix can guide a researcher to certain protocols so that they can develop and expect certain results from the research activities chosen.

Examples of generative research tools include collages, card sorting, diagramming or mapping, directed drawing or sketching, physical modeling with premade building blocks such as Velcro shapes, Legos, or similar. Human subjects are provided tools to allow them to express latent needs, unexpressed aspirations, and essentially be a co-participant in the design process. The term co-design referenced by Elizabeth Sanders, PhD, alludes to the new capability and skill set that designers possess. The ability to use a multi-method approach to understanding, learning, and forecasting that is directly relevant and contextual to the participant, provides a powerful process that leads to new and innovative thinking and output. Sanders describes the three general research methods as what people say, what people do, and what people make (Fig. 2.4, Sanders (2001)). Corresponding disciplines are listed to denote what field takes full advantage of studying these human behaviors. Business marketing is very effective in tools that elicit, what people say. These include interviews, questionnaires, and surveys. The field of anthropology uses various methods and tools for observation and recording human behavior to track what people do. Psychology and Design have adopted the third domain as a method for understanding and working in concert with people.

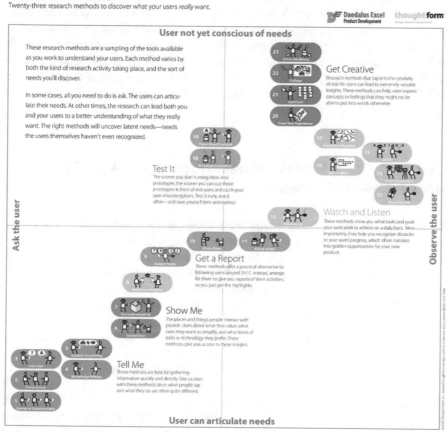

Fig. 2.3 Research toolbox, Daedelus design, thoughtform

Fig. 2.4 Say, do, make framework, sanders

When using a range and number of methods, designers gain a broad and deep understanding of people, their needs, and their unexpressed aspirations. The underlying belief is that people have a wealth of knowledge and valuable experiences that can lead to critical insights. And designers use and create their own custom methods and tools to draw out a participant's latent needs and desires. The following case study illustrates the use and outcomes of various research methods to gain unique insights into an everyday appliance and seemingly mundane task of microwave heating and cooking.

Case Study: New Product Market Discovery by Applying Human-Centered Research

Whirlpool Corporation collaborated with undergraduate junior level, Third-year Industrial Design and Visual Communication Design studios to tackle a saturated market, the microwave oven. The North American market buys over nine million microwave ovens annually (Statista, 2017). With highly competitive overseas manufacturers, the challenge is uncovering new markets. Cost can only be optimized to a diminishing return. So how does a company reimagine a rectilinear box that heats food with a given technology? The teams could have easily approached this as a redesign and styling project. However, sustained market expansion can only take hold if the innovation matches the market and intended use and users. Designers are no longer seen or relegate themselves as repackaging stylists. Rather than starting in the right side of the double diamond process, the teams develop research strategies that investigate how people use this method of cooking in their everyday life. Madsbjerg and Rasmussen's book, "The Moment of Clarity" separates everyday phenomena such as travel, eating, exercise, and banking—which can be analyzed using property data points from the hard sciences or experiential aspects from phenomenology (Madsbjerg & Rasmussen, 2014). If biological gender (man or woman) is a property, then cultural gender (masculine or feminine) is the aspect. Science can help us determine if a person is a man or a woman, but how do we find out what it means to experience masculinity or femininity? What is being a man or woman like? Only the study of the phenomenon will help us understand that. Phenomenology is the study of how people experience life.

It is the entering an everyday space with completely fresh and unsullied impressions to understand a routine activity so that we can decipher deeply ingrained personal rationale and actions. Thrust into a behavioral researcher role, designers use their creative capacity to develop research protocols that glean the human aspects of preparing a meal with a given technology. The design team visited participants' home and conducted an introductory interview in their living room. After a few questions, they placed a blank piece of paper and pen in front of them. The designers asked the participants to redraw the interface of their microwave. The intent of this simple exercise was to ascertain frequently used buttons. All participants sketch a numeric keypad, an instant 1-min start button, and start and stop/cancel buttons, but never explicitly draw specialty buttons. There can be more than 12 highly specific buttons meant for fish, popcorn, dinner plate, defrost, etc. Immediately, this exercise allowed the design researchers to quickly understand the manner in which participants typically use and remember their microwave oven use. The follow-up observational research at participants' kitchen provided more unique anecdotes typically inaccessible through alternate methods. The participants were observed to press the buttons in this sequence: "5, 5, Start" or "2, 2, 2, Start" as their method for invoking 0:55 s or 2 min and 22 s for cook times. It seems that participants were creating their own shortcuts by not moving their fingers to other numbers. Some of the design researchers were amused, and some shocked at the perceived "laziness." But as the research of people's habits and ingrained rationale of food preparation with a microwave oven was uncovered, they saw unique opportunities. From an interface standpoint, the ramifications for one team's design criteria were direct interactions. This focus was built on discussions between using the microwave oven and the conventional baking oven. Most subjects would never cook a full meal in a microwave. Whereas using the oven typically requires a bevy of planning, preparation, selection of particular pans, and presentation. In comparison, microwave oven users consigned their device to simple reheating and singular servings. These are the aspects of the behavior dictated by how people use their technology in relation to their environment, activities, and other people they are serving. Examples of the newly proposed designs spanned the uppercase Design, strategic market shifts, and lowercase design, as design form and interfaces directly addressing the participants' lifestyle and usage of the microwave oven. Figure 2.5 shows a range of products developed to reach particular market segments. In top left to right order: Space Saving model was developed for the person who eats and reheats out of Tupperware containers. By understanding single people, the teams observed that they would make primary meals and save them in a plastic container storage for future use. Instead of plating the food, the participants would eat directly out of the plastic container and return it to the refrigerator. The second example, Personality and Character model was in response to subjects commenting on the how ugly their microwave ovens looked. Subjects spoke of their "FSS" flat surface syndrome problem in the kitchen. Wherever there was available space, they utilized the space. This compounded the original problem as food packages and other kitchen stuff was changing and hiding the appliance, since all microwave ovens have a horizontal surface. A subtle curve on top build in a design affordance discouraging articles to be stored on top; the use of feet supports and door face strengthen the personality and character of what is seen as an inanimate object. The intent is to have the microwave oven as a foreground object that owners take pride in, rather than a background object. The third example is in response to vehicle travelers who use microwave ovens in their motorhome recreational vehicles, 18-wheeler truck cabins, boaters, and similar. Designed to be bracket mounted so the entire unit does not shift, and internal slots for custom mugs and bowls with lids, was only discovered through ethnographic and contextual research methods. The last example was

Fig. 2.5 Whirlpool microwave oven prototype models

developed for public or shared spaces such as cafeteria or common eating areas. Simplified controls and interactions were the main thematic design criteria.

The Studio as a Tangible Place for Active and Agile Learning: Space and Time Matter

Studio-trained designers are inculcated through a unique way of working. This work occurs in dedicated spaces that afford thinking, processing, and reflecting on the problem at hand. This recursive awareness for the interaction between person and object is part of the mental exercise that happens on a constant basis. Added to this mix are peer colleagues who are part of the informal and formal conversation. Their resident knowledge and experiences benefit the studio cohort. This hive of activity, work, and discussion allows for a unique and rich learning environment. This section is about the combination of factors that dynamically create an active learning and reflective experience. The key factors include the studio environment, studio mates, time, and critiques. Each of these distinct elements is in service of human-centered mindset.

The School of Design's Product Design track philosophy is based on the interaction between the person and the object and the system created by these relationships. This mantra is taught early and constantly through the curriculum. The course structure referenced (Chap. 5, Figs. 5.2 and 5.3: SOD Curriculum) shows the once students have beginning skills in Form and Context, their Sophomore Year second semester is explicitly Design for Interactions. Each design studio in first and second years is coupled with a design lab. The studio lab concentrates on the technical skills, machines, and software required for the trade. The design studio sets the stage for what and how to approach a problem. The paired studio and lab courses work in tandem to build off each other's content. Analogous to the Making and Doing quadrant, the activities and output mirror the intent of learning by actively experimenting and reflecting. In contrast to other visual and performing art fields such as a music conservatory method or an apprentice model, the design studio + lab model is a mix of teacher and peers working and learning together. The one-to-one student–teacher ratio is unsustainable from several points of view. To carry the music analogy forward, the design studio is more similar to an informal orchestra or band constantly working together. The individual practicing scales and shifting is the equivalent of sketching and low fidelity modeling. Once individual skills have reached a competent level, team projects become more frequent. Solo success is de-emphasized. The sum of the parts should be greater than the whole. And the equivalent of a musical jam session in longitudinal and visual form produces unique and interesting experiences that could have never been preplanned.

The engineering education model of lecture and lab is similar in intent. The main concept to be learned is presented in a large lecture hall with examples and case studies. At a later time and date, students attend smaller class lab. If there is no hands-on work, a recitation would be equivalent. A subtle but significant difference in engineering recitation and lab is the assigned instructor. In most engineering cases, a Teaching Assistant (TA) or Graduate Assistant (GA) is the primary instructor. The quality and experience may vary significantly from year to year and class to class. On the other hand, product design studios are almost always staffed by a full-time professor, not a graduate student. If there is a TA or GA, they are in support of the design professor and act as a secondary voice. Just as all assistants, they can vary in both teaching and professional experience. Thankfully, the design studio structure has the professor/s as the primary instructors. This structure is predicated on a larger mechanism though. One of these includes the predominant number and scale of each respective discipline. At the time of this publishing, Carnegie Mellon University's College of Engineering reports 1780 full-time undergraduates, 1850 graduate students, and 184 faculty members. The School of Design student numbers were 180 undergraduates, 50 graduates, and 33 faculty members. Also engineering sponsors and funds a significant number of doctorate and PhD students. Some are dedicated to research and may not see the inside of a classroom. But the structure of an academic unit is predicated on delivering courses to an x number of students every year. And these Masters and PhD students factor into the teaching quota.

Though the design field and CMU School of Design do have PhD and DDes degrees offered, the number of these students are less than a dozen. And typically half will teach into a lecture (non-studio) design course or elective courses.

Modalities of Working, Thinking, and Reflecting: Environment Supports Learning

Imagine a study group formed for one of your college classes. Often, these groups are self-formed because of friendships. Other times, they are formed out of a practical need to have more brain power and different points of views aimed at complex problems. Now imagine that group is 12–20 of your fellow classmates in a dedicated room who stay with you as you matriculate through freshman to senior graduation. This is a classic industrial design studio arrangement. A typical industrial design studio class admits 10–30 students. If an institution admits a larger cohort, then additional sections are added. But there is always a studio space created as a place for student work, lectures, presentations, and everything in between. This is similar to professional businesses, where employers will have open work spaces with dedicated seating. This type of physical space, studio cannot be understated. Though one can trace this type of work space to the Bauhaus, its precedent is the co-op artist studios, the apprentice model, and professional studio offices. The benefit of working in this manner has many direct and indirect outcomes, especially for a student. The shared studio becomes a place for learning and playing while building social camaraderie. It literally becomes a community of practice.

One of the prevailing philosophy for this type of arrangement is a design student will learn as much from your peers as their professors. When given an academic home in which to play, discover, fail and have fun, students are learning constantly in formal and informal ways. The power of place and space coupled with longitudinal "living" with one another creates an environment of co-learning and co-teaching. Every year, an engineer working with a product design student will come to the studio space and the look of wonderment is obvious. They understand what a unique work atmosphere a studio affords. Most engineering programs do not have a practical nor feasible manner of creating individual private desks that are accessible 24 h, 7 days a week to a large number of students. Lawson and Dorst's (2009) book, Design Expertise lists five factors that studio environments provide as co-location, learning-by-doing, continuous access, integrative learning, and mimicking practice. Keeping these in mind, this chapter and latter Chap. 5 expands on these concepts in practical descriptions and examples.

The physical space allows work to be shown and have a common home. As much as digital work, cloud-based collaboration, and remote work is useful from a productivity standpoint, working synchronously is critical. The advantages to having project work in at least one physical place afford: communication of ideas, sharing of the work, process of work, collaboration on the work, and visual evidence of decisions and/or progress. This is a very dry and antiseptic description of an environment that provides more than a place for stuff. But, nothing can yet compete with face-to-face interactions. Highly collaborative professions or parts of a project that require direct and immediate interactions cannot be easily replaced with asynchronous workflows. Though corporations and professional environments are collaborating regularly, there is always some form of face-to-face (either in-person or with technology) that enables team members to focus on a common problem at the same time. As a learning environment, the design studio arrangement is a powerful stage that allows many elements to work together in instructional, prescriptive, and reflective ways. And as an educational platform, it becomes a type of training ground for young designers who graduate to professional environments where a combination of in-person interactions and digital collaborations occur.

Design thinking has many definitions. But in the case of this text, design thinking can be considered a way of approaching and working on a problem by diverging and converging processes with the human stakeholders as its focus. And design thinking is enabled and enhanced through the work spaces, time constructs, and like-minded people. A place which a student considers their own permanent design home for each semester (and sometimes the entire academic year) enables play, discovery, teamwork, and collaboration. The permanent studio environment (as opposed to hot-desk studios where students bring and take all their work and supplies to and from residence to school) cannot be discounted nor minimized for its multifaceted advantages. The greatest benefit is the organizational and behavioral effects of working, living, and playing with a common cohort or design family. Figure 2.6 is used as a catch-all working space, play space, meeting place, classroom, and presentation space. Well before corporate offices adopted this type of environment, educational and professional design studios were working like this from their inception. Leveraging multiple perspectives and the ability to work collaboratively in the same space affords common team focus. Classically trained studio designers who experienced this type of education realize during and post educational experience that they learned as much from their peers as from the instructors. The learning is not necessarily subject specific to the problem sets but is the lighting stick for relevant second- and third-degree subjects surrounding the project. These related subject areas are offshoots and inquiries into how things work, how people act or behave, how are things made, cultural aspects, and other fundamental questions. This subset of problem-solving builds students who are independently resourceful investigators conducting their own literature searches. This part of digging past the surface and delving into the technological, economic, social, political, and relevant factors is part of the macro and micro points of view a designer undertakes to build a convincing argument to their proposed design solution.

Combine a dedicated space with a project that requires several months of dedicated work, begets a student who will have considered many points of view. Independent learning in areas related to the project requires the student to be proactive and take on a role such as a private detective or research scientist. They are after reasons and causes and the relationships between similar and dissimilar things. Students generate a range and depth of proposed solutions in various mediums to simulate a future state. This future state could be simple aesthetic considerations. Or

Fig. 2.6 Design studio space examples

it could be a method for testing how a new object may invoke different interactions between person and machine. And all of these sub-problem solving activities are about exhausting as many scenarios they could build within time and resource limitations. Not unlike an advanced Reggio Emilia approach, the design studio and project challenges set up conditions for self-challenge and motivation. A layered support system of people, place, and time, enables an atmosphere of play and exper-imentation. The studio becomes a place on its own—it creates its own social and cultural atmosphere. Implicit and explicit, formal and informal learning happens throughout the studio experience.

Project-Based Learning and Open-Ended Problem Sets

The studio space is a reflection of the type of work being assigned. The intermediate and advanced problem sets are project-based rather than exercises that have predict-able results. The National Science Foundation, Accreditation Board for Engineering and Technology (ABET) have called for changes in the engineering education (NSF, 2007). Some of these changes are driven by external factors and input from industry. Other factors are in relation to prescribed learning objectives and goals. These calls are not necessarily new, rather even more critical as the complexity of large-scale projects and their impact on the society and natural world is increased. Clive Dym's et al.'s journal paper, "Engineering Design Thinking, Teaching, and Learning" (Dym, Agogino, Eris, Frey, & Le, 2005) supports the use of project-based learning, PBL as an effective educational method for teaching design. Though they are in the camp of engineering, the perspective of a designer is universal in the sense that the practice is of "systematic, intelligent process in which designers generate, evaluate, and specify concepts for devices, systems, or processes whose form and function achieve clients' objectives or users' needs while satisfying a specified set of con-straints." They go further by highlighting the skills associated with good designers.

- Tolerate ambiguity that shows up in viewing design as an inquiry or as an itera-tive loop of divergent–convergent thinking.
- Maintain sight of the big picture by including systems thinking and systems design.
- Handle uncertainty.
- Make decisions.
- Think as part of a team in a social process.
- Think and communicate in the several languages of design.

These bullet points closely resemble Nigel Cross' seminal book "Designerly Ways of Knowing" (Cross, 2007) five aspects of identifying a person who may pos-sess designerly ways of knowing:

- Designers tackle ill-defined problems.
- Their mode of problem-solving is solution-focused.
- Their mode of thinking is constructive.
- They use "codes" that translate abstract requirements into concrete objects.
- They use codes to both read and write in object languages.

Cross posits that design education should be a third part to the studies of science and the humanities. If the sciences study the natural world, the humanities study the human experience, then design study is about the artificial world. The traditional science and humanity studies use a combination of empiricism and rationalism to build a model of the world. Each study is steeped, for good reason, in its historical methods and approaches to seeing the world. But when it comes to building new parts of the world, the world of the artificial, a hybrid and new approaches are necessary.

Open-ended challenges force students to develop and build something that relates to particular criteria. The studio project brief is created as an open-ended challenge for students to self-determine their trajectory. They set the bar by identifying potential areas that can be researched through the exploratory and generative research methods. The project evolves by the diverging and converging paths that build towards the domain goal. This domain goal can be characterized as a high-level problem framing. Chapter 3 unpacks framing and reframing the context in more detail. And Chap. 6 expands fully on the value of introducing open-ended problem sets early in the curriculum.

Essentially, design criteria are typically not fixed until the last possible moment. Rather it emerges from the behavioral and ethnographic research findings. And as any company or project manager knows, their product systems is highly contextual. There can be clear specifications at the onset of every project, but acknowledging and allowing shifts in these in relation to incoming research is essential. By recognizing this, the modality of work results in thinking and acting differently. If we refer back to Ullman's Design Paradox (Fig. 1.4), it is the designer's freedom and flexibility that gives them a unique advantage over an entrenched employee that has worked in the same industry for years. F. Scott Fitzgerald's sentiment that "… holding two opposing ideas in the mind at the same time and still retain the ability to function" represents a unique type of intelligence. Similarly, the industrial design discipline believes and practices that holding two or more realities in the mind is the norm. The designer is constantly imagining and reimagining ideas, interactions, and experiences by creating assets and props to simulate the proposed future reality.

Human-centered designers enter through the human condition. The ability to ask seemingly obvious questions that get to the cognitive essence of behavior creates a new point of view that can produce insights and the next innovations. As a designer attempts to understand the subject matter in these ways, they can derive contextually appropriate criteria in which to address the main problem and sub-problems. Working through this type of dynamic process with human guiding principles starts distinguishing a designer from other disciplines. The gap in the ability to imagine and propose new realities widen because of the tools, methods, and skills studio designers possess. This significant difference is further highlighted in Chaps. 3 and 4. The combination of the designer mindset, tradecraft, and use of the collaborative iterative process, provides a powerful repertoire to tackle simple to wicked problems.

Methods for Continual Introspection: Design Studio Critique

The design critique is a method combining presentation and critical review of how a designer's work in relation to the design brief. The design studio space not only facilitates this formally but provides a safe space and time for practicing between peers informally. The critique is an essential part of being a reflective and critical designer. To simplify into two categories, there can be formal critiques or informal critiques. A formal critique is when the student presents and defends a body of work they have created in response to a project brief. This event when the instructor and/ or guest subject matter experts ask the reasons and rationale for various decisions made by the students. These can encompass everything about the final form, material choices, manufacturing processes to the viability of the proposed product as a better solution for the intended users and market. Formal critiques typically have a time limit and the student has carefully curated the visual work and verbal presentation. Each student within the class presents their work and receives some form of critical feedback from instructors, SMEs, and student peers.

What prepares students for this Socratic methodology of critical dialogue to uncover presumptions, new or better ideas, and spark reflection is the informal critique. Informal critiques occur between instructor and students and students and students. The student to student critiques is always occurring inside or outside of class times. This is where each person is sparring and honing their mental and verbal defense of their decisions. The time during and post critique allows the student to reconsider their decisions and approaches to a problem. Do I need another round of testing? How is this idea more competitive than what is currently on the market? Is just changing the aesthetics differentiates my design from other brands? Existentially, is my proposal for making millions of these widgets environmentally responsible? As long as the critique is forcing the student to reflect upon their processes that generated their current project state, then the format and purpose of the critique is doing its job. The informal critique happens daily with peer to peer interactions. The frequency of informal critiques varies with instructor to student. However, the studio course instructors typically meet with students on a biweekly basis or more frequently in the North American structure. Thus some form of ongoing discussion is part of the design process. This usually mitigates any final formal critique surprises. Lastly, in-person formal and informal critiques require synchronous timing. There is no substitute for being in the same space at the same time, focusing the discussion on the work immediately in front of the audience. As simple as this sounds, aligning the grounding philosophy of actions and decisions are self-evident through this tried and true method.

The formal critique can be a very stressful experience for a design student not used to defending their work. It is most harrowing when they are unprepared or did not consider very basic facets of the project statement. Some instructors can be ruthless. Sometimes it is for effect, and sometimes it exposes obvious flaws in the student's judgment. And part of these misjudgments is due to the lack of understanding the essential context of person to product interactions within a larger world view.

Fig. 2.7 Design studio critique, Lyons, J. (2018). Carnegie Mellon University

Design critiques are not for the weak. Nor is an attitude of I'm in the right. The critique is just another foible in which to consider high-level uppercase Design decisions that can be affected by functional, aesthetic, ergonomic, design decisions. It is a reflection tool that cuts through the newly proposed design and alternate reality.

When a student transitions from the educational environment to the professional, the critique is replaced by other terms such as reviews and presentations. Instead of the professor's obtuse questions, designers and engineers take lead and responsibility for their desired direction. One of the main purposes of the educational critique is for test the student's conviction and viability of the idea. The second part, the viability of manufacture, functional integrity, etc. is relatively rote and can be learned. Once a student grasps the basic tenets of how things work, design for manufacture and assembly, cost and benefit analysis, and other business and technology principles, the challenge becomes defending their case for the qualitative results. Time and time again, product adoption can come down to meeting a combination of business, technological, and emotional aspects. And the empathy given to an inanimate object creates the tipping point for the highly desirable and most coveted goods. The qualitative element can be technically proficient and can become the deciding factor between one version of a consumer good and another. Thus, the ability to have the vocabulary, narrative, and self-conviction behind a product's highly qualitative reason for bringing it into existence is paramount. The newly minted designer or engineer will use various forms of acumen to assert their point of view for every new product they intend on generating. And the critique is the starting point of finding their way and methods for proposing and defending each of these projects (Figs. 2.7 and 2.8).

Fig. 2.8 Design studio open critique, Finkenaur, J. (2016). Carnegie Mellon University

Engineering Excels in Evaluative Stages

Engineering education excels at evaluating what has become before them. These can include prior inventions and products, case studies of processes that led to discoveries, and accidents or disasters to learn from earlier errors. The reasons and benefits of this process are multifaceted. The underlying principles of an engineering curriculum are applying empirical knowledge in mathematics and science to create and test ideas. As obvious as this may sound, it is the sequence and lack of courses that engineers are exposed to early in their education that determines a particular mindset and ability to approach problems.

Each discipline uses historical references to learn and build from. However, each curriculum and course applies these quite differently. For example, if we were to develop a new cordless power drill, a patent search to an engineering student would be to understand the mechanisms, electrical requirements, gearing ratios, bit holding method, manufacturing and assembly of distinct parts and systems, etc. A designer's purpose would be concerned with how to allow both precision and power hand grip for a tool to transfer energy into another material. Additionally, they would want to address the repeated use and misuse of this tool and see how others before them defined a particular solution. It is acknowledged that both disciplines would have nearly the same intent and purposes as they read over prior patents. However, the priority and mindset of each discipline can and has been observed to be very different. Subsequent to this literature and background research process, it becomes even clearer that each discipline has different priorities. Engineering students will look for ways to optimize, create efficiencies in material and cost reductions, whereas a product designer will zoom out and try and reframe how to best accommodate the operator. The reframing of the project brief typically entails an understanding the bigger picture first, then proposing new paradigm shifts to solving the problem. Creativity, seeing things differently, and exploration is imprinted early onto the mindset of studio-trained designers. Details on the methods for shifting perspective to gain both detailed and meta-level aspects on a product development follow in both Chaps. 3 and 4.

Earlier in this chapter, the description of open-ended challenges set the stage for a certain way of processing and thinking. As such, these types of problems require a different set of methods. And these methods are built for discovery and a dynamic, generative workflow. The early core classes of engineering education typically do not rely on project-based courses, collaborative opportunities, and provide highly challenging scenarios. The ability to imprint early in the educational growth trajectory has passed. Several schools and institutions have recognized this need though. Coined as cornerstone courses, these project-based learning classes present open-ended challenges in the first years of academia. But the major disconnects are faculty who can teach this type of pedagogy, the environment and team culture that allows for failure and exploration, and a grading or reward system based on soft-skills and process rather than production.

Carl Sagan (1996) said "Science is more than a body of knowledge. It's a way of thinking." Sagan's concern of humankind's grasp of world knowledge in the age of science and technology requires more than just deep investigation. Also, it requires critical thinking and creativity to solve seemingly simple to complex problems. And having designers and engineers as proponents of the human experience requires a planned composition of educational structures and delivery. An agile thinking mindset requires a combination of curriculum, environment, and timing: like-minded cohort of peers who support each other as they matriculate through the curriculum, properly devised challenges which allow the students to find their own methods and processes for discovery and inquiry, spaces that support discussion, debate, and discovery, and a system of instructors who can deliver accordingly. Enabling young designers and engineers to be agile in thinking, processes, and production is the heart of problem-solving, innovation, and progress.

References

The American Heritage® Medical Dictionary. (2007). Retrieved April, 2018, from https://medical-dictionary.thefreedictionary.com/scientific+research.

Cross, N. (2007). *Designerly ways of knowing (Board of International Research in Design). Board of International Research in Design (Book 100)* (1st ed.). Basel: Birkhäuser Architecture.

Dym, C. L., Agogino, A. M., Eris, O., Frey, D. D., & Le, L. J. (2005). Engineering design thinking, teaching, and learning. *Journal of Engineering Education, 94*, 103–120.

Explorable.com. (2018). Retrieved February, 2018, from https://explorable.com/definition-of-research.

Hanington, B. (2007). Generative research in design education. *IASDR07 Conference, Emerging Trends in Design Research*. Hong Kong Polytechnic University, November 12–15, 2007.

Lawson, B., & Dorst, K. (2009). *Design expertise*. London: Taylor and Francis.

Madsbjerg, C., & Rasmussen, M. B. (2014). *The moment of clarity: Using the human sciences to solve your toughest business problems*. Boston, MA: Harvard Business Review Press.

National Science Foundation. (2007). *Board for Engineering and Technology (ABET) have called for changes in the engineering education*. Retrieved from https://www.nsf.gov/pubs/2007/nsb07122/nsb07122.pdf.

Sagan, C. (1996). *Science is more than a body of knowledge. It's a way of thinking*. Carl Sagan's last interview.

Sanders, E. B. -N. (2001). Virtuosos of the experience domain. In *Proceedings of the 2001 IDSA Education Conference*.

The Statistics Portal. (2017). Retrieved March, 2018, from https://www.statista.com/statistics/220122/unit-shipments-of-microwave-oven/.

Chapter 3
The Praxis of Design: Framing, Making, Doing, and Defining

Design Processing: A Composite of Activities and Outputs

When starting a new project, product designers conduct several activities to holistically understand the current state of the problem space. Some of the first activities include gaining as much insight from the client, identifying stakeholders for future research, and literature searches. The next stages become a whirlwind of thinking, discussing, doing, making, and learning, all in context of the intended project.

Even though there are many known design processes which abstractly describe the general progression, the linear and inherently serial aspect of the visualization does a disservice to the actual actions. Having taught and worked for over 20 years in the product design and development field with Fortune 500 companies ranging from commodities to high tech electronics to automotive manufacturers, I can confidently say that the simplicity of the double diamond process or any linear process is not indicative of the detailed and nuanced thinking, activities, and outputs of a designer or engineer. There is a distinct need to educate beyond the novice designer with a more representative visualization of the dynamic actions and output that occur in project-based learning.

Significant research from mainly architecture, computer science, and engineering disciplines have proposed various taxonomies to dissect the design process' inherent complexities into descriptive and prescriptive methods. Finger and Dixon's (1989) paper Review of Research in Mechanical Engineering Design does a comprehensive analysis of other authors and their own assertions. The references speak to the highly social aspects and the inherent value of the interactions between designer, the act of designing, and their environments (Bucciarelli, 1988). Also, it is Stults' (1987) point of view that design is a fractal-like process where the act of designing repeats at varied times and different levels of detail. The challenge to describing something in generalities and find the common patterns removes the explicit nature of what is trying to be expressed.This chapter is a response to the overly antiseptic nature or too high-level description of the process that does not enable students to plan, act, and

© Springer Nature Switzerland AG 2019
W. C. Chung, *The Praxis of Product Design in Collaboration with Engineering*,
https://doi.org/10.1007/978-3-319-95501-8_3

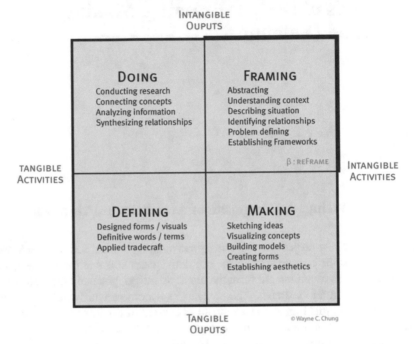

Fig. 3.1 Design praxis matrix

reflect upon a flexible guide. This chapter proposes a dire need to better represent and provide a visual format and method to convey the nuanced aspects of designing and contextually varied types of activities. Ultimately, the subsequent matrix is aimed to be practical and prescriptive while providing reflection when relevant to the student designer.

The revisualization of the design process is shown as a 2 × 2 matrix (Fig. 3.1). These highly iterative processes are categorized and identified as Framing, Making, Doing, and Defining. Each of these quadrants of the design process has varied activities and outputs. And each of these has corresponding tangible and intangible activities and outputs. One of the main reasons for re-visualizing the design process is because most projects do not follow a clean or linear path. The Design Praxis Matrix borrows from the business management professors, Ikujiro Nonaka and Hirotaka Takeuchi. Their SECI matrix (Nonaka & Takeuchi, 1995) (socialization, externalization, combination, and internalization in relation to organizational behavior) is a visualization of knowledge transfer and social interaction between individual employees and an organization. Nonaka and Takeuchi use two descriptors, tacit and explicit, which define distinct behavior attributes and actions to each quadrant. The Design Praxis Matrix is built in a similar fashion with Tangible and Intangible as the primary concept descriptors; and Activities and Outputs as the modifiers. These concept descriptors relate to each quadrant's main descriptors (Framing, Making,

Doing, and Defining) associated with the complex and dynamic design processes. The following section defines the quadrants primary attributes. However, a fuller descriptive discussion follows to enable application and theory to coincide.

The Framing quadrant is in layman's term, getting your head around the overall problem to establish context and relationships between essential elements. The designer searches for information in relation to the intended problem area. They are typically using third-person or literature searches and any prior subject matter information to gain an overall sense of the project and problem background. Framing is the designer's interpretation of the overall context of the project. However, as the design process is highly iterative, a reframing occurs once new information and knowledge are synthesized. And the point of view and approach on how to address issues re-adjust accordingly.

The Making quadrant is working in tangible form to represent ideas, concepts, systems. It is the process of visualizing ideas, conducting first-person research, connecting ideas and concepts, and analyzing and synthesizing information to make relevant sense of and for conceptual relationships. These representations are relatively low-fidelity or lower resolution than a finished model. This closely resembles converging actions and output in relation to the double diamond process. This is the compilation of applying an individual designer's own experiences, current mindset, and external world views to activities and converting these as outputs for both internal team members, stakeholders of the project, and the external clients.

The Doing quadrant consists of activities and outputs that connect the designer to the primary humans in context and corresponding stakeholders. Performing behavioral exploratory research, generative research, and evaluative research is explicitly enabling empathy to be part of the design process. The purpose of these activities is to gather the information so that a constructive and collaborative phase for analyzing and synthesizing contextual information can occur. The output from this quadrant spans back and forth with the Making quadrant as these dynamic quantitative and qualitative activities require output for reiteration and results.

The Defining quadrant is the manner in which a designer and engineer will objectify the solution. This is a tangible, high-resolution, clear representation of what the designer imagines as the solution to the stated problem. Activities and output in this quadrant are physically tangible because both can typically be observed. Creating a highly specified physical or digital model has explicit steps and procedures that represent the training or education of the designer. The praxis and practicing process of analog sketching, computer modeling, physical modeling, and other experiential skills are self-evident.

This revisualization of the overall design process enables understanding of the dynamic cognitive and physical activities that generate particular outputs. Rather than a simplistic linear visualization of the design process, the Design Praxis Matrix more closely represents the unpredictable nature of the design project processes and design thinking. A spiral movement or a back and forth flight between several quadrants occur because of the highly contextual results of the thinking and doing. The activity of thinking equates most closely to the intangibles, whereas the behavior and output of doing is evidenced in tangible artifacts and activities. This matrix captures the varied scales of the design process' theory and practice. Thus, reframing (focus and refocusing of the problem or situation), an intensive iterative process, is more easily understood and better demonstrates how this occurs in relation to a team's behaviors. It is important to explicitly depict varied activities and outputs are not strict predecessors on a linear progression. Rather they are adjacent and highly relational activities and outputs. The additional benefit to both inexperienced and experienced teams is removing the long linear and serial concept of completing one stage of the design process to move to the next. Other diagrams will note or show that the linear process can have concurrent actions also happening. As much as this may be representative and unrepresentative of the concept, these visualizations do not take into account the actual actions involved, nor the non-serial progression that occurs. And the Design Praxis Matrix affords concurrent or multiple activities and outputs to exist amongst several quadrants. And lastly, the matrix also recognizes that design practice can enter into any one of these quadrants. If a project is a redesign version 2.x, then there is minimal need to abstract concepts and reframe the earlier defined criteria and project focus.

An analogous manner to imagine the diverse roles of this matrix is building a house. The idea or concept of building a particular style, size, etc. residential living space equates to the Framing quadrant. Framing or roughing in with the structural members and other lattice represents the general understanding and intent of the dwelling. The Making and Doing quadrant is akin to hanging the drywall, running electrical and plumbing and adding insulation. These acts and outcomes are usually hidden by the final interior finishing, but are necessary for a modern house's function. The craftsmanship of details and making it your home requires Defining and personal choices. Each quadrant is necessary and requires a different setup, skills, and knowledge. To provide a product-specific example of the Design Praxis Matrix' in action, a traditional redesign problem of a cordless power drill is described below in Fig. 3.2.

Earlier in Chap. 2, the reference to Nigel Cross using the term "codes" is understood as translating abstract requirements into concrete objects (Cross, 2007). This is in deference to Christopher Alexander's, A Pattern Language: Towns, Buildings, Construction book (Alexander et al., 1977). Alexander identifies a pattern constructing activity as a fundamental design process. To find or build a pattern, a language is necessary. Thus Cross use's code as the building blocks for a constructive process. These code building blocks are the equivalent of the tangible and visual tools described in more detail in Chap. 4 as concept maps, diagrams, sketches, and models. So in relation to the Design Praxis Matrix, designers are building their own specific project language because of the contextually unique code representing the research findings.

Agile and Fluid Frameworks for D/design

The Design Praxis Matrix becomes a powerful and flexible tool for design teams. The Defining quadrant's activities and output are highly tangible since it is the literal aspects of aesthetics, ergonomics, form giving, material choice, and manufacturing

Design Praxis Matrix

Intangible Ouputs

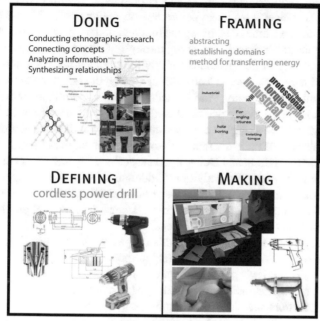

	DOING	FRAMING	
Tangible Activities	Conducting ethnographic research Connecting concepts Analyzing information Synthesizing relationships	abstracting establishing domains method for transferring energy	Intangible Activities
	DEFINING cordless power drill	MAKING	

Tangible Ouputs

Fig. 3.2 Design praxis matrix example

considerations. The actions in this quadrant require mastery of 2D and 3D design tools, skills, and techniques—thus the applied tradecraft phrase. The members of the team have highly defined forms and terms to describe their solution, and they are building tangible and visual constructs of a digital or physical object. The lower right quadrant Defining is diametrically positioned in relation to Framing. Framing requires little to no experience to engage in the design thinking processes. This is the inherent power and inclusiveness of the quadrant. However, to create actionable and significant change, some level of expertise beyond general making is necessary. A sense of craft, experience in wielding the right tools, manipulation of the tool, jig making, sensemaking, and other studio-trained and professional experience is essential to capitalizing on the lower left axis quadrant.

The upper left quadrant Framing consists of many individual and team cognitive activities and abstractions to define the problem space. This quadrant typically includes traditional behavioral research methods to understand people and users in current and past contexts. Framing activities use representations of tangible objects through words, phrases, stories, or scenarios to define the situation. Post-it note activities commonly populate this quadrant. The less tangible outputs are the phe-nomenological understanding of the people, place, and products of the project space. This quadrant enables teams to agree on terminology, mental models, and consensus of the problem and/or opportunities. This is the beginnings of Cross' constructive process and Alexander's language between team members. In Fig. 4.2 power drill example, the design team is describing the fundamental purposes of a cordless power drill to ascertain the end goals. By going through this process, the team may arrive at a new technology or new methods for driving threaded fasteners into a material. During one of these exercises, a team commented on the fact that homeowners have a drill to mount framed photos. And immediately others mentioned that solving this problem is now more easily achieved through 3M wall hooks. Even though the design team may be working for a power tool company, it does not exclude the abil-ity to think laterally or build alternate solutions to shift the paradigm.

The quadrants, Making and Doing is a dance of activity and output that straddles both tangible and intangible activities. Evidence of this work is apparent when entering a product design studio. The wake of material, adhesives, tools, fasteners, paint, and other collateral is part of this multi-staged, incremental problem-solving actions that aid next decision-making steps.

It has been observed that the quadrant's output can be characterized as internal facing or external facing output. Abstracting output is typically internal to the design team as to designate their own common denominator or baseline criteria. Rarely does a team present abstract words and images to clients because it is undefined in relation to project/product document requirements. However, the Defining quadrant obviously satisfies the external facing attribute to clients, whereas the Framing and Making quadrants can both be internal and external tools for aligning the design team and/or communicating the process and benchmarks to the client.

Design process divergence is most notable in the Framing quadrant. Convergence has to occur in the Defining quadrant. And both divergence and convergence occur between Framing and Making, and this will sometimes feed back to the Framing quadrant and can be considered Reframing. There can always be vertical or horizon-tal movements while working in the matrix framework. The vertical combination of Making and Framing are used when generative and participatory research methods are being conducted. Generative research methods require building tools or props to get people to participate in the design process. The repertoire of Making and Reframing is a natural outcome due to the directed activities and output.

The actions, output, and dynamic thinking that results from one another neces-sitate a descriptive term: "D/design." As mentioned in the earlier chapter, Design thinking is accessible by all and is represented by the uppercase Design activities. However, studio-trained designers implicitly and naturally work at both lowercase design levels and metacognitive Design levels to fully leverage the discipline. D/design is meant to describe the complexity of utilizing multiple modes of design to address a problem or undefined situation. Using the nomenclature d/design signifies Defining and tradecraft skills; whereas D/Design signifies high-level, paradigm-shifting processes and intent. These undefined situations are better known as wicked problems. Coined by professors of urban planning Rittel and Webber, they describe these problems not in degree of difficulty, but as issues so unique in comparison to the norm. And therefore prior known processes cannot resolve them. Examples of these wicked problems are climate change, food waste, changing migratory habits of animal species, or access to clean water. The nature of these problems requires more than a single solution or myopic treatment of a dynamically changing systemic situation. Everything from policy planning to choice of building material has an effect on addressing parts of the unsolvable problem. The flexibility of the design process is seen as a template for synthesizing knowledge and people in ways other disciplines have not been able or intended. Designers and design thinking thrive when collaborating with others.

The Design Praxis Matrix as Force Multiplier

The unique value of this matrix is the ability to enter, move, and iterate through the structure for a team's own purpose. Product designers have so aptly used this process from the Framing quadrant and working back and forth movement between Making and Doing quadrants because it allows them to find meaning between all of the variables and actors. This meaning builds a common ground that can be explained in the human terms because the research is driven from this point of view. Ultimately, it is easier to make a story and subsequent decisions when solution criteria are based on human needs, wants, and rationale. Ultimately designers end or pass through the Defining quadrant where highly designed digital and/or physical objects are built. And sometimes in this quadrant, the team realizes they still did not achieve their own utility or aesthetic standards. This calls for reentry into the matrix to rectify the issues depending on time and resources. However, depending on the project intent, a design team may enter into any one of the four quadrants. This is the first force multiplying factor: the point of entry into the matrix is the dealer's choice. Most non-designer projects utilize three of the four quadrants: Framing, Doing, and Making. Non-designers or teams without a medium in which they can craft with high-fidelity or resolution cannot perform in the Defining stages as described in the prior text. However, this does not disqualify all others who are not traditionally trained designers, engineers or technical fields. If one has their own tradecraft where material or medium can be manipulated to represent a concrete concept, then this can be considered their own language of Defining. However, as previously mentioned, the power and flexibility of design thinking allow non-designers to apply the majority of the activities and produce output that is relevant and useful to their context. The ability to enter into the matrix at any angle allows the managers, leaders, and participants know their roles and expectations before moving to a new quadrant.

The second multiplying factor is the movement through the matrix. The prior examples all have described a relatively contained sequence that starts with Framing the problem, diverging and abstracting the situation, to Making concepts and Doing/conducting research protocols, to Defining a highly designed solution. However, there are times that call for moving against the grain in atypical movements diagonally or across quadrants. If we use a traditional product design scenario where the intent is an incremental modification to an existing product, that team may enter at the Framing quadrant and move immediately to the Defining quadrant. The reason may be simply lack of resources such as time and funding. Or it could be a regular product cycle revision with highly experienced team members who do not need to Reframe the situation. Instead of seeking radical change, the design thinking process is still used but particular activities and outcomes are not necessary. And in this case, explicit and controlled management of trajectory is planned and executed. As the reader can see, the term force multiplier is appropriate since the quadrants are more than just a sum of their parts. And as users of the Design Praxis Matrix apply each quadrant, significant outcomes occur through the constructive dynamic outputs and novel options are available in an adjacent quadrant.

The β beta symbol placed outside the Framing quadrant represents additional layers as design teams iterate and build upon successive rounds of design thinking and making processes. This is an advantage of multiplying factors inherent to the Design Praxis Matrix: iterations on many planes. Design teams can use the prior layers of the matrix as both a guide and history of decision mappings. These can be useful for student reflection or professional review of a project.

Lastly, the Design Praxis Matrix has inherent value through its visual design. Without having a structured format, non-designers and designers alike may not be cognizant of their approach or path. It is essential to know the limits of particular processes so not to have unrealistic expectations. And it is critical for managers and leaders to recognize the differences of what (quadrant factors) and how (entry into the matrix, movement through the matrix, and iterations through the matrix) the design thinking process can be maximized and leveraged in their organization.

Acknowledging Other Design Processes

There are many established design processes and principles prior to this revisualization. These are still inherently relevant and necessary for explaining the design process. However, many of the principles can be applied to individual quadrants and/or from a systems perspective. We can use the accepted diverge and converge descriptions when describing movement and actions in most of the quadrants. Most importantly, if we identify empathy as an essential criterion in which a design team needs to consider and embody, the activities and output of quadrants Framing and Doing would lend itself to this fundamental need. These quadrants' properties are most concerned with tangible activities such as mixed methods research; and intangible outcomes would be considered findings, patterns, and themes from the research. However, the same emphasis can be aimed at any of the quadrants resulting in linkages and commonality of empathy throughout. If we situate ourselves at the stage of Making with empathy as a driving criterion, team members aim activities and output that are sensitive and responsive to the human participants. Other established design thinking principles such as define, ideate, prototype, test, and collaborate can all be applied either directly to a particular quadrant or globally to the entire spectrum of quadrant factors.

It has been observed that the quadrant's output can be characterized as internal facing or external facing output. Abstracting is typically all internal output that is useful for the design team to designate their own common denominator. Rarely does a team present abstract words and images to clients because it is ill or undefined in relation to project/product document requirements. However, the Defining quadrant obviously satisfies the external facing attribute to clients, whereas the Framing and Making quadrants can both be internal and external tools for aligning the design team and communicating the process and benchmarks to the client.

This matrix may not perfectly fit every design project's demands. However, having used this implicitly and explicitly in various projects of complexity, it satisfies many levels of designers and engineers. And the main intent of this author is to educate and build self-direction and reflection into the design process.

Reimagining Old and New Ways of Working

This chapter shares a simple but powerful model of how creative people move through a process. This representation can satisfy a range of people and teams from the professional to first-year, freshman students. Saul Bass said: "Design is thinking, made visual." And having a visual tool to orient people to the nebulous activities and outputs of design thinking is necessary. Product designers are habitually concerned with material, weight, finish and methods for manufacture. To achieve this level of specificity, there is an ongoing need of sensemaking within their processes. Sensemaking can be considered a process by which people give meaning to experience. For product designers, they use tactility, appearance, and approachability of the object. They use the activities and are sensemaking within the matrix to achieve these results. For non-designers, it is important that they are self-aware of what they are looking for and how to make sense of the activities and objects that they will produce. Otherwise, many designers move through a design thinking process blind to the intent and value of the output or activities they are performing. Subsequently, if a user of design thinking has the knowledge and confidence of the journey and tools necessary, self-agency occurs. The Design Praxis Matrix is a compliment to highly experienced designers and design thinkers. They are outcome oriented in implicit ways that an outsider may not perceive. For non-designers, the Design Praxis Matrix is a roadmapping generative tool. Rather than a serial path, the matrix allows for flexibility in process and thinking. It is easy to digest but has an inherent potential for complexity. And empowered through this tool, non-designers and designers alike become evangelists for the robust process.

Ashleigh Axios (2017) succinctly states Design is recognized to have an impact across a range and depth of areas. https://designobserver.com/feature/to-be-a-design-led-company/39637/

She sums up the strategic values in three points:

Design as a tactical driver: where design alters a discrete product, service, or, communication effort

Design for system innovation: where design alters an existing system or creates a new one to deliver a better solution

Design as a catalyst for transformation: where design changes attitudes and behaviors of a community or organization

This chapter's intent is to show that the design process needs to be imagined and is being leveraged in many different ways. The Design Praxis Matrix is a step towards better representing the power and dynamic of a highly adaptive process. This process is both a tool and reflection on the people applying it. Design is and has been recognized as a tool and process for tactical and strategic thinking. As companies constantly seek to innovate and imagine more than higher efficiencies, but a new way of doing business, the Design Praxis Matrix distillation begins to allow for this meta-level and micro level of understanding and control. The following case study for Ford Motor Company illustrates the d/Design activities, output, and thinking required to tackle complex and undefined problems or situations.

Case Study: Ford Motor Company's Future of Mobility

The future of mobility is affecting our society, economy, and the environment in profound ways. As we recognize at a global level, constant growth and consumption continue at unsustainable rates. Under the influences of urbanization, 70% of people will live in cities by 2050 (Steer, 2015), increasing global prosperity where people will own 2.5 billion cars by 2050 (UNICEF, 2012), morphing markets, and the way we perceive and recognize mobility will drastically change. Broadly speaking, mobility is the accessibility for people, goods, and services to go where they need or want safely, efficiently, and affordably. However, business as usual is no longer the norm in the automotive industry. Technology in both physical and digital medium is affecting how business is planned, managed, and executed. Therefore, new ways of thinking and imagining one's core business is necessary to remain competitive and innovative.

So how does a traditional automotive manufacturing company imagine what is possible in the near and far future? The Ford Motor Company, Human Machine Interaction Lab partnered with the Carnegie Mellon University's School of Design to seek and develop new business services and systems for the North American market. Ford's HMI Lab recognizes that the human experience is paramount to delivering successful products and services. They also recognized that disruptive innovations are occurring in the marketplace that would have never been conceived without certain technology and infrastructure. Businesses such as Uber, Lyft, Zipcar, and PeaPod are changing how people and goods are moved. These are enabled through a combination of app-specific software that tasks service providers to move in a particular physical mode of transportation to a requested location. As simple as this quotient may seem, ridesharing services were created from using underutilized resources (Eckhardt & Bardhi, 2015). Others may argue that these are examples of a sharing economy relying on social networks. Regardless of the actual definition, transportation corporations recognize that these very different business models are in direct or indirect competition. It is difficult ignoring companies that are being valued in the multi-billions of dollars within 7 years of their inception. And even if one does not perceive it as immediate competition, there is recognition for new market opportunities in the mobility of people, goods, and services.

The project initiative from Ford Motor Company HMI Lab was to apply user-centered methods for framing new mobility opportunities. And this definition of mobility was not primarily concerned with the manufacturing of traditional vehicles. The broad questions included what are the domains that will make a difference in mobility? What type of technology and systems of technology will make an ideal user experience? How does lifestyle dictate what types of mobility options a person may choose? How can mobility be affordable economically, environmentally, and socially? How can the experience be simplified, seamless, and fun?

Interdisciplinary Project Overview

Eight interdisciplinary teams of consisted of at least three members. And each team had at least two designers, a communication designer, and an industrial designer. The third or fourth member included computer science, human–computer interaction, engineering, architecture, or humanities majors. The course is a longitudinal mixed method research project based studio. Significant emphasis is given to applying traditional behavioral research methods that help develop participatory, generative action tool methods. Concurrent with the research processes, design teams use visual conceptual mapping and modeling to understand the problem space. Progressively in the process, design teams create low fidelity prototypes and/or generative tools to test and iterate with participants. Depending on the design opportunity, the teams design physical and/or digital solutions. These designed artifacts act as assets for a future state scenario that culminates into a final video sketch. The caveat to this paper is a tremendous amount of time, experience, range and depth of skills were present in each team. This paper cannot fully represent each team's efforts or design processes. By including images and diagrams inline, the reader can hopefully understand the depth in which these teams performed.

Significant challenges occur when managing interdisciplinary teams. Concerns for enabling each member to utilize their own experiences and capabilities are paramount. Organizational behavior can easily sidetrack the purpose of the project, or teams may not even be on the same track. One of the common denominators used to align team member actions and outputs is the design thinking process.

When presented with the relatively broad scope of the project, teams devised research plans that would give them a holistic understanding of transportation in the North American market, everyday situations that require movement of people or goods, and future trends related to mobility. Several topics immediately gravitated towards the emerging changes facing the current infrastructure's inability to cope with the rising population's needs. The final topics of each of the eight teams included: multi-modal transportation needs, onboard driving assistance for new drivers or impaired drivers, a system and service for moving your house contents between residences, sharing tools and devices like electronic goods within a community, leveraging the interior vehicle space for mental health, priming students for a successful learning experience on the school bus, multimodal transportation modes, and gamifying the commute to relieve stress.

Design Praxis Matrix's Application and Utility

To demonstrate the flexibility and use of the Design Praxis Matrix, this paper will describe how teams used this framework to make sense of a boundless problem space. At the project start, one particular team named ARGO entered the Framing quadrant. The Design Praxis Matrix's internal quadrant descriptors portray the tangible activities involved as problem defining and finding relationships between users and current mobility issues. The ARGO team's first activities included literature searches and research to frame the problem space. The intangible outputs included a range of text statements and problem statements that framed the areas of opportunity. Output consisted of relevant parts of business reports, initial first-person and third-person research findings, and concept representations of the problem space. Within this quadrant's activities and outputs were a series of searching, finding, and refining/redefining the queries to narrow the problem's scope. The team saw significant opportunities addressing the movement of goods from one location to another. This initial stage of the process provided the team members an understanding of the past and current states of moving services, products, systems, and stakeholders in a metropolitan city.

Next, the team used the Framing quadrant to zoom out to ensure they are seeing the current problem in its entirety. The types of activities and outputs are primarily intangible. These include brainstorming words, terms, and concepts that enable broad thinking of the problem space. Example output of the activities is typically created on whiteboards and/or post-it notes. Though this quadrant may constitute simplistic or minimal processes, the activities can play a pivotal role in the other quadrant's progress because it can be seen as the point of inception for great ideas. An example of this is shown in Fig. 3.3.

With these abstracted elements, the team fully defined their problem statement within the Framing quadrant. In this case, they set out to address: Household moving services in the USA lack fluent communication with users and fail to provide successful moving experiences. With current and future technology, moving services can be much more efficient and provide a sense of trust and security to end-users.

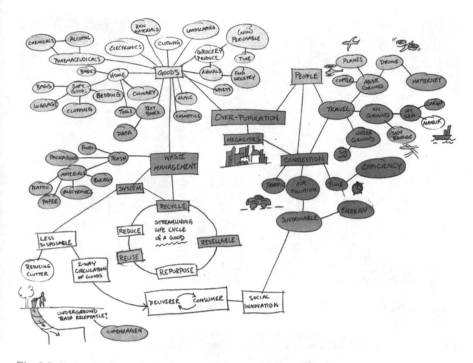

Fig. 3.3 Brainstorming mobility futures. Aislinn, T. (2014). Carnegie Mellon University

Interestingly, both Framing and Doing quadrants were used as they developed protocols for traditional and generative research methods. Specifically, in the Formulating quadrant, the team bridges and combines various technologies to an antiquated business: residential moving services. Developing concept maps provide a basis for understanding the current moving services and systems. The ARGO team's future scenario was a combination of 3D scanning, smart tape for box contents and tracking, printing technology that embeds electronic information specific to the box contents, and syncing information to the mover and end-users device to be displayed throughout the moving and unpacking process. The team's name stems from the acronym Automatic Remote Graphics Operation. See Figs. 3.4 and 3.5.

These stages consist of highly dynamic back and forth combination of activities and outputs. This fluid process is less of a recirculation of events as presented in prior design process diagrams. But more of an unbounded play between each sub-goal and task at hand. Texts that reference intuition and the magical moments of design (Kolko, 2015) are better represented by the matrix's type of movements throughout the design process. It is less about a repetition of steps, but more about harmonizing relevant activities and outputs of each design thinking quadrant to scaffold to the next appropriate action or output. That moment can be about activities that support better understanding or outputs that lead to better decision-making. This case example happened to primarily portray the Framing, Making, and Doing quadrants. Brainstorming sessions that use the Framing quadrant can leverage the

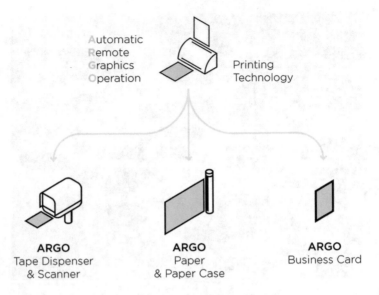

Fig. 3.4 ARGO overview, Lee, J. (2014). Carnegie Mellon University

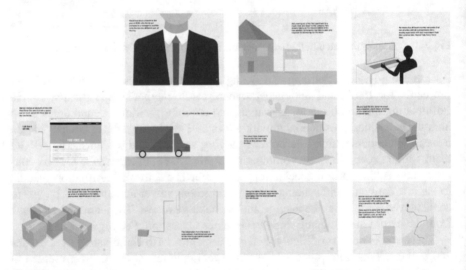

Fig. 3.5 ARGO storyboard scenario, Kang, J. (2014). Carnegie Mellon University

Making quadrant for examples of materials that lead to new product application insights. For example, when teams are introduced to the material properties of gorilla glass, light-emitting metal, or flexible interfaces, their preconceived notions of what is possible is suspended. Suspension of prior beliefs allows the Framing quadrant's activities to move with different momentum and directions.

Each team comprised of a communication designer, industrial designer, and one or more non-design specific discipline. This complement of students is relatively

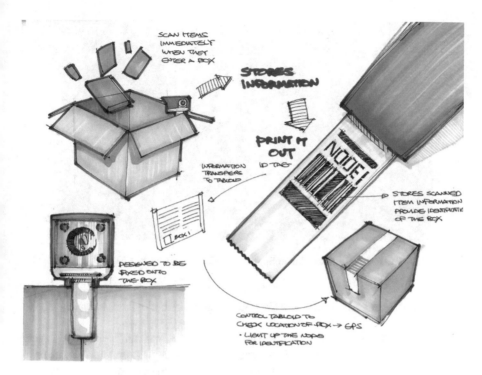

Fig. 3.6 Concept development forming and framing quadrant outputs. Lee, J. (2014). Carnegie Mellon University

important. Addressing nebulous, unbounded, open-ended problems requires a diverse group with various experiences, perspectives, and skill sets. The ARGO team realized that to deliver a better future of residential moving services, they would need both physical and digital products working in concert. Specific product artifacts and consideration for material technology were possible through the combination of team knowledge. See Fig. 3.6 for the range of idea concepts developed during the Making quadrant as a direct response to the Doing quadrant's research analysis.

Resource limitations such as time, budget, or personnel require producing a solution in response to the original problem statement. Inevitably, teams move toward the Making quadrant where activities and outputs are both tangible. These include concrete design solutions, visual and physical artifacts that require discipline-specific skills and activities. In the case of this design course, the physical products and digital assets become props for a final video (Fig. 3.7). The ARGO team and two others were selected to represent the ongoing innovative research being conducted with Ford Motor Company. https://social.ford.com/content/ford-social/en/articles/technology/what-does-the-future-of-mobility-look-like-to-millennials-.html

The story was picked up by Fast Company's online periodical fastcodesign.com. The following video link is the future scenario envisioned by the design team:

Fig. 3.7 ARGO moving service video clip. Lee, J., Kang, J., Yang, H. (2014). Carnegie Mellon University

http://www.fastcodesign.com/3046268/moving-students-devise-a-smarter-way-to-pack-boxes

The Ford Mobility project required students to participate and perform in several design and research-led spaces. These spaces are the inception and core to the product design, service design, and/or user experience design solutions. The competencies required from students include the ability to be empathetic, enable others to be part of the generative dialogue to frame the problem and be part of the design solutions. The Design Praxis Matrix allows design teams to use their expertise and prior research, while keeping an open conduit to the constituents through exploratory, generative, and evaluative methods. Eventually, the design team has to commit and conduct activities and output in the Making and Doing quadrants, by definition of their role as designers. Consequently, all of these activities and outputs become stepping stones to a longer-term vision.

References

Alexander, C., Ishikawa, S., Silverstein, M., Jacobson, M., Fiksdahl-King, I., & Angel, S. (1977). *A pattern language: Towns, buildings, construction, series: Center for Environmental Structure (Book 2)*. New York, NY: Oxford University Press.

Axios, A. (2017). *To be a design-led company*. Retrieved from https://designobserver.com/feature/to-be-a-design-led-company/39637/.

Bucciarelli, L. L. (1988). An ethnographic perspective on engineering design. *Design Studies, 9*(3), 159.

Cross, N. (2007). *Designerly ways of knowing (Board of International Research in Design). Board of International Research in Design (Book 100)* (1st ed.). Basel: Birkhäuser Architecture.

Eckhardt, G., & Bardhi, F. (2015). *The sharing economy isn't about sharing at all*. Harvard Business Review.

Finger, S., & Dixon, J. R. (1989). Review of research in mechanical engineering design. Part I: Descriptive, prescriptive, and computer-based models. *Research in Engineering Design, 1*, 51.

Kolko, J. (2015). *Exposing the magic of design: A practitioner's guide to the methods and theory of synthesis, Human technology interaction series*. New York, NY: Oxford University Press.

Nonaka, I., & Takeuchi, H. (1995). *The knowledge-creating company: How Japanese companies create the dynamics of innovation*. New York, NY: Oxford University Press.

Steer, A. (2015). World Resources Institute. Retrieved from http://www.wri.org/blog/2015/12/transforming-our-cities-cities-we-need.

Stults, R. (1987). *Computer graphics 1987: Proceedings of CG International '87*. Tokyo: Springer.

UNICEF. (2012). *An Urban World Report*, Retrieved from http://www.unicef.org/sowc2012/pdfs/SOWC-2012-An-Urban-World-map.pdf.

Chapter 4
Methods for Building Future States: Framing and Reframing

Mapping The Current State to Reframe a Preferred State

We have seen business, industry, and society, change quite dramatically with new technologies and business models. Everyday transportation has been transformed by Uber's ability to access underutilized resources. Some may say this type of social innovation allows anybody with a car to provide on-demand transportation services. Regardless of how to categorize or define the company, Uber has built a new model for mobility. Between late 2015 and late 2017, Uber's valuation fluctuated between $48 billion and $68 billion. Bloomberg (2017) This is for a company established only 8 years ago in 2009. To put some perspective on this, General Motors was founded over 107 years ago and its valuation was $50 billion as of May 2017. Forbes (2017) Who knows what the future may hold. It is quite possible that the astronomical ascent may come crashing just as quickly. Regardless, the distributed access to many services through smart devices and cellular and other communication infrastructure have allowed opportunities for Uber, Lyft, and Airbnb to exist. Business models have and will continue to be flipped and new paradigms emerge. Add the increased use and integration of machine learning and artificial intelligence, the landscape of everyday work and life are going to be altered in disruptive manners not yet imagined. Company giants such as Facebook, Amazon, Netflix, and Google will be affecting some part of our lives in very direct manners. And if not directly, there will be indirect ripples that will course correct competing or alternate industries that they encounter.

As technology develops, so does the manner in which humans and machines interact as a system. Product design is not just about tangible objects. They have roles as form givers, researchers, interaction designers, service designers, social innovators, and strategists. How can a single profession that is humbly termed "product design" place so many people in such diverse roles who are prepared for so many different responsibilities? Why have so many industries turned their

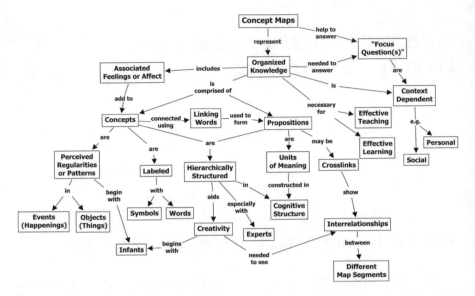

Fig. 4.1 Concept mapping explained, Joseph D. Novak and Alberto J. Cañas

attention to design's magic methods for innovation? What makes the process so unique and valuable? Working in both the current world and imagining an alternate one requires a range and depth of known knowledge coupled with tacit and explicit simulations. Product designers are required to forcsee and forecast future products, environments, contexts, and human behaviors in order to make appropriate design decisions. Also, designers are futurists regardless if they asked for the job title and implied responsibility.

Various visual and collaborative tools allow both mapping the current state and projecting for a future state. Concept mapping is a critical tool for understanding everything from simple products to complex systems. By visually organizing the main concepts of a system, teams are able to understand and discuss the relationships between each concept and the corresponding linkages. Collaborative processing and communication between internal and external parties benefit from building a contextual common denominator. There are several types of mapping methods. We use the concept map as a grouping of specific terms to visualize relationships. As a whole, the map describes a central idea or system. The combination of Hugh Dubberly's (2017) methodology for building concept maps and Novak and Cañas (2017) Fig. 4.1 illustrates the overall concept, purpose, and benefits. Building one essentially is forming sentences or prepositional phrases. The concept is a noun connected by a linking verb or preposition to another noun/object. Noun, verb, noun'. The trailing noun' links to other objects. A simplified rule guide provides a consistent output which enables designers and engineers of all levels to capitalize on this collaborative tool:

- Title of your product/system/service; reframe it by zooming out
- Label all links and concepts
- Concepts appear once

There can be exceptions depending on the purpose of the map and the intent of the team. Sometimes the trailing noun is an attribute or adjective. If this occurs, that typically ends the tree branch. Other exceptions may include stating more than one user. This may occur if the map is describing an interaction or many stakeholders. And in these cases, each user is given a unique term such as sender and receiver or primary stakeholder and secondary stakeholder.

Product designers and engineers use concept mapping for several reasons. Among the advantages is working visually so that all team members can determine discrete elements of the system. The low-tech version of paper-based, post-it notes inherently allows teamwork to exist. As mentioned in Chap. 2, working in the same space at the same time allows for internal teamwork. Developing a means to discuss concrete to abstract articles of the given product system enables joint understanding. This common denominator is essential to create a starting point to move forward the discussion. By creating this tangible output, team members are able to define the overall purpose of the system so that successive maps and conceptual models can be developed. This mode of self-management enables collaborative behavior amongst team members.

The following example, Fig. 4.2 shows a traditional representation of a physical appliance, a sewing machine. Note the domain name and the subsequent reframing of the purpose of a sewing machine as a method for connecting soft goods. If we were in the business of attaching pieces of soft goods together, then our brainstorming and ideation phase could propose glue, staples, tape, fusing, and many other connecting methods. However, if we were in the business of manufacturing sewing machines and were looking to improve on the effectiveness of our current method, the concept map allows the team to determine the relationships between objects or the actions that allow failures to occur when operating the device. If you have used a sewing machine before or ask someone who previously sewed, one of the repeating issues is that the thread breaks. There are many possible reasons why this can happen. But using the concept map allows the team to identify that the tension through the take-up lever pulling from the bobbin is one of the main culprits for this failure. The team may reevaluate how to dynamically adjust the tension through variable mechanisms between the bobbin and take-up lever and thread guide assembly. And maybe the team may reexamine whether they can control all types of thread material and their tensile strength. And at this point is the design, engineering, and business decision could be about controlling the system by providing and selling their own type of threads that are tested and will not ruin the sewing experience. Just by visually capturing the overall device's individual elements, designers and engineers are able to reframe in either left or right sides of the double diamond by using several matrix quadrants.

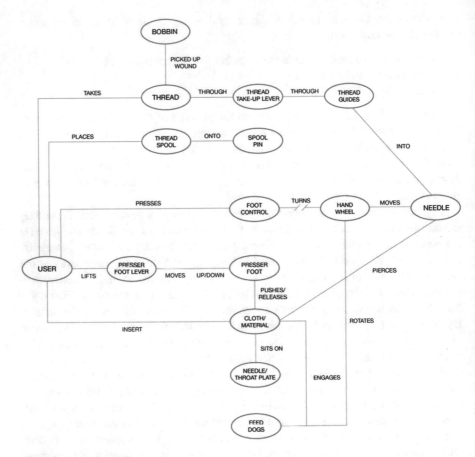

Fig. 4.2 Sewing machine concept mapping

And this repertoire is only possible because the team members have depicted the device as they perceive it in a tangible form for everyone to discuss.

However, the reader saw a glimpse of how concept mapping can provide a broad view of nontangible objects such as service or system in Chap. 3's Fig. 3.3. Pulling back the veil of the process shows that concept mapping is a highly dynamic and analog activity. Figure 4.3 is a result of countless changes to the noun objects post-it notes, whiteboard links, and descriptor verbs or prepositions.

This version of concept mapping is not to be confused with website mapping or site mapping. The intent is to describe individual concept entity in relation to inter-action. So in the case of a website, the touchpoints or clickable links spider out as a strict hierarchy map of the system. Rather, if we plot out the relationships of Instagram the website as a concept map, we would portray Fig. 4.4. The overarching concept, of what Instagram does as a business model can be represented. Imagine what a Pinterest mapping may look like? Surprisingly, these are not overly complex

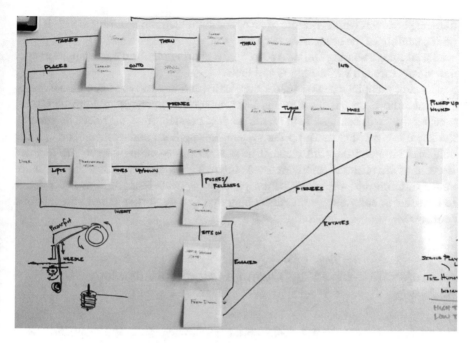

Fig. 4.3 Sewing machine concept mapping analog process

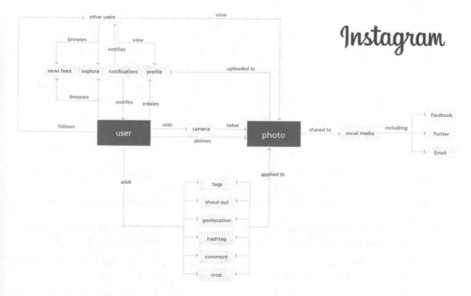

Fig. 4.4 Instagram concept mapping

mappings. How it works, is more of an operational diagram resulting in a multilevel understanding of the company's purpose.

Social Image Sharing Online App/A digital sharing platform that changes images and videos through various filters and tools.

Concept mappings can be used strategically and tactically for qualitative analysis. Once the team has the systems level perspective, they are able to determine whether they will continue in the same product domain. If so, they have deciphered the key elements that make up the system. Discussions and decisions on specific concept terms and their interrelated links and corresponding objects are enabled because of the prior team activities and output. This is can be another tool for innovative thinking and designing. Concept mappings inherent medium and relative low-barrier to entry allows interdisciplinary people to work together in a common framework.

Appropriate Technology Through the Phenomenology of People

From Chaps. 1 and 2, the way designers enter a problem space is through the lens of people. Because of this approach and philosophy, design students are taught to gain an understanding of the person in context. Phenomenology is the study of the structures of experience and consciousness. As a philosophical movement, it was founded in the early years of the twentieth century by Edmund Husserl and was later expanded upon many students and philosophers (Gallagher & Zahavi, 2012). Existential and hermeneutic philosophers such as Jean-Paul Sartre and Martin Heidegger were among the many who criticized and gave form to the movement. The philosophical movement that "seeks through systematic reflection to determine the essential properties and structures of experience" (Menon, Sinha, & Sreekantan, 2014). There are several assumptions behind phenomenology that help explain its foundations:

> Phenomenologists reject the concept of objective research. They prefer grouping assumptions through a process called phenomenological epoché.
> They believe that analyzing daily human behavior can provide one with a greater understanding of nature.
> They assert that persons should be explored. This is because persons can be understood through the unique ways they reflect the society they live in.
> Phenomenologists prefer to gather "capta" or conscious experience rather than traditional data.
> They consider phenomenology to be oriented towards discovery, and therefore they gather research using methods that are far less restrictive than in other sciences (Orbe, 2009).

Studying people in situ is necessary because each person has had their own individual experiences. Because of this, they act and react to objects, the environment, and other people in unique ways. The best product designers and engineers are ones who are critical observers. Critical in what type of material evokes confidence and trust between person and product. And critical observers of other people in situ and why they interact the way they do. Designers are constantly probing why a person acted the way they did to determine the reason behind their behavior. What makes a person tick? The resulting value can be overarching contextual principles which define design aesthetics and lowercase design aspects. Thus, designers see everyday people as the experts in their own right. To convert their phenomenological understandings, a constant activity and output of concept mapping, synthesis, and analysis transpire throughout the design process. To provide a glimpse of this rich internal team activities, output, and insights, a case study of the exploratory and generative research process follows in both verbal and visual details.

Case Study: Cognizant Technologies: Understanding Human-Centered Activities and Output for Contextual Discovery

The following case study highlights the combination of various research methods and mappings that directly affect the subsequent concept directions, and ultimately determined the AR technology design solution. The praxis of understanding through people, visualizing and mapping the knowledge, analysis, and synthesis to build new methods and tools for deeper discovery, and a repertoire of making, thinking, and reflecting is represented in the following series of images and descriptions of a three-person interdisciplinary team.

Carnegie Mellon University's School of Design partnered with Cognizant Technology Solutions on a semester-long elective studio project course called UX Design Tools. The intent was broad in its inception: identify emergent opportunities where technology will play a significant role in people's interactions and experiences. What is the future of physical space? How are advancements in IoT, augmented reality, and telematics influencing how we experience environments?

This project's premise is at the heart of today's role of technology and how people will adopt and interact with the world. Currently, people are reliant on a range of personal technology products to being networked into a complex digital ecosystem connecting work, play, and everything in between. With logarithmic advances in computing power, artificial intelligence, new material technology and manufacturing options, design and designers are at the forefront of humanizing these tools and interactions.

What processes can be employed to leverage digital, material, and unimagined technologies? Discovering these possibilities and how technology may play an essential role in our new future is why ongoing investigations through contextual research are essential. And it is a necessary tool to mine the diverse range and depth of each business and industry sector's opportunities. The possibilities to change and improve people's everyday lives by understanding humans before technology open greater possibilities for impact. And designers are trained to seek and see things through the lens of the user. Designers possess a unique set of abilities that generate new perspectives and unique opportunities by using behavioral research skills, analytical abilities, collaborative nature, and generative skill sets. They apply these methods and skills in both intangible and tangible building to propose and produce unrealized solutions.

In the perpetual march of exponential technological progress is important to note that this project's approach purposefully started with observing and engaging people, as people and not as users. Human-centered designing starts with a particular mindset. This mindset requires empathy and understanding an individual to societal levels. Being contextually aware and listening to other people's experiences is the starting point and essence of qualitative research richness. We have seen prior examples of when man–machine interfaces were not properly considered. Or when efficiency is prioritized over human needs and control is stripped from a human operator. It is the study of people and their behaviors that is the core enabler of how latent opportunities are derived. This mindset and journey of discovery provide a map to natural and safe solutions to the human–machine interface—and ultimately ethical technology development centered on human needs.

Seven interdisciplinary teams with three students per team were organized in a 15-week elective course called UX Design Tools. The students ranged from junior, senior and graduate level students. Each team consisted of three members, comprised of at least two designers, a communication designer, and a product designer. The third member included computer science, human–computer interaction, engineering, architecture, or humanities majors. The course is a longitudinal mixed method research project based studio. Significant emphasis is given to applying traditional behavioral research methods and ethnographic methods early in the course. These phenomenological aspects may become the basis for developing participatory, generative tool methods.

Concurrent with the research processes, design teams used visual conceptual mapping and modeling to understand the problem space. Progressively in the process, design teams created low fidelity prototypes and/or generative tools to test and iterate with participants. Depending on the design opportunity, the teams designed physical and/or digital solutions. These designed artifacts act as assets for a future-state scenario that culminates into a final video sketch. Each team member has overlapping levels of experience, range and depth of skill, and some prior internship or professional experience. This paper highlights portions of the teams' processes and artifacts created that lead to deeper human understanding and ultimately their product and service solutions.

Cognizant is one of the world's leading professional services companies, transforming clients' business, operating and technology models for the digital era. Their industry-based, consultative approach helps clients envision, build and run more innovative and efficient businesses. Headquartered in the U.S., Cognizant is ranked 230 on the Fortune 500 and is consistently listed among the most admired companies in the world. Their specialties include Analytics and Information Management, Business Process Services, Intelligent Products and Services, CRM, SCM, IoT, Cloud, Intelligent Automation, Infrastructure Services, Quality Engineering and Assurance, Application Services, Enterprise Risk and Security, etc. As a large-scale, multinational technology service provider of information technology, consulting, and business process outsourcing services, Cognizant is aimed at identifying and developing new solutions, services, and strategies that change the way we experience the world and do business.

To be a continual leader in this highly competitive field, Cognizant focuses on being a partner for end-to-end digital transformation—believing a successful enterprise must understand the context in which they compete and invest in new business, operational and technology solutions. To commit significant investment towards transformation, Cognizant needs a robust understanding of the primary benefits and goals to the business and intended audiences. One of their means of staying at the leading edge is contextual research—a combination and series of ethnographic and longitudinal research uncover problems or gaps through the investigation and process of human interaction and behaviors.

Team Compass: Self-Election Behaviors and Interactions

Compass team started simply by attempting to help students with mental disabilities. Several weeks of literature reviews and interviews with trained educators in special needs helped them further focus their project scope. The main qualifier to this course is that students have to have access to the targeted participants in order to administer first-person interviews and subsequent action research. This criterion ruled out a significant portion of mental health subjects due to Institutional Review Board regulations and course limitations. However, college students with attention deficit hyperactivity disorder (ADHD) self-identified and were willing participants in the project. The unique aspect of this participant group was they conquered part of their own disorder in their own way. Their presence in a higher-education academic institution proved to themselves and others, that they can achieve goals regardless of their personal disabilities. The team focused their research process and methods on six, self-identified ADHD college students. One of the primary takeaways that served to be a difficult challenge was that ADHD students all learn differently and there is no "best practice." A second mantra was simply stated: "being successful in school is a lot about getting stuff done." Interviewing these students was incredibly valuable because they had already developed their own tools—self-awareness, cognitive strategies, time management, and other methods of accomplishing short-term tasks that built towards larger goals. With numerous organizational tools in both physical and digital forms, the team used these prior tools as a basis for mapping and understanding basic fundamentals and benefits of each one. Alternately, from the human perspective, they identified that productive people needed: flexibility, focus, and motivation. Contextually to ADHD college students, Flexibility meant the ability to manage or flex to a task dynamically and not build strict rules or timetables; Focus was the ability to be notified or shown relevant tasks in relation to their location and other task priorities; and Motivation was a personally relevant and perpetual way of showing progress and improvement.

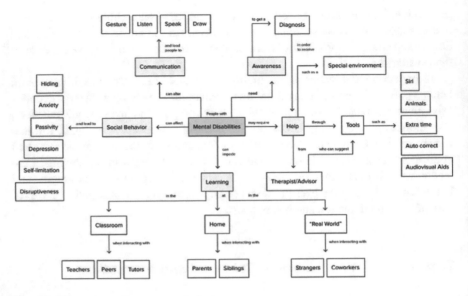

Fig. 4.5 Concept mapping compass project

The following images and illustrations will provide an overview of the research methods and processes this team employed to arrive at their mixed augmented reality design solution. As with most design thinking and design processes, there is a convergence, divergence flow to the process. Few texts provide an inside look at the raw and personal nature of the process. The following series of edited figures representing diverging and converging activities and output presents the related image with short explanations.

The team's literature review and early interviews with educators led to a broad scoping of the potential situations surrounding mental learning disabilities. Figure 4.5 is an initial step to framing the problem space.

At this moment in the research process, since the discussion was about technology, the team used provocations to elicit future ideas and scenarios where the participants could imagine themselves. This exercise coupled analog mapping of a four-quadrant grid set up by semantic differential terms related to current scheduling and their digital equivalents (Fig. 4.6).

Figures 4.7 and 4.8 represent visual activities to get participant reaction to free-hand line drawings depicting current ways people manage and schedule work. These are examples props that are not purposely un-finished in form so that they enable participants to impart their own personal experiences and stories. Interestingly, a moment of divergence occurred when the team analyzed and started mapping the responses from the ADHD participants. If the tools and techniques were so useful to this population, why could not others benefit from them? Additionally, most of the tools were not ADHD specific. So how were others using these tools and how are they different from the intended user group?

Fig. 4.6 Exploring existing digital learning environments research

Figures 4.9 and 4.10 portray the daily activities in relation to a participant's functional and emotional aspects. Understanding and designing for appropriate user experience require multidimensional aspects of physical, behavioral, emotional, and cognitive considerations.

Figure 4.11 is a natural extension of visually organizing and presenting a typical day with the purpose of finding the "curve balls" or incidents that would easy affect an ADHD student from their intended regimen. The added dimension of time in this visualization method is a necessity to how the participants are using actual time in relation to perceived and planned time.

Figure 4.12 converged on the cognitive specifics of how the participants would currently plan their day. The figure shows the participant imagining how a calendar and to-do list may work if it was considered Dynamic Scheduling. By asking them to move out of their comfort zone and reimagine a new way of working is an effective generative, future-casting method. This resulted in new ideas but some friction on how it may be delivered.

Fig. 4.7 Initial concept provocation cards

Figure 4.13 is the team's need to test earlier findings against a universal population. Limited to six ADHD participants, the team needed to ensure a broad perspective was not lost by engaging 24 participants. This diverging but confirming generative research method reinforced their initial findings, but also helped galvanize specific themes and criteria for the design solution. As expected, the divergence stage progressed to a convergence of design criteria. The activity was a sheet that asked: "To manage my time better, I wish I had (fill in the blank)." The activity sheet then asked participants to draw and illustrate their answers.

User Category Themes originated directly from generative tool user responses. This part of the process was transformative towards key themes that defined the tangible design concept. Highlighted themes that were expressed included: Removing Stimuli, Self-reflection, Feedback, Peer Support, Rewards, Motivation, Dynamic Scheduling, Prioritization, Health, and Physical Space. In realistic terms, one product may not be able to address all of these inclusively. So a subsequent remapping of an intended user's day was graphed to these high-level themes to

Fig. 4.8 Initial concept provocation exercise

conceptualize which features and interactions may be best addressed with near-future technology (Fig. 4.14 is indicative of the method and responses). Figures 4.15 and 4.16 represent a synthesis and convergence of understanding how the participant group perceives and aspires to manage their world.

Subsequent figures focus and convergence towards a final tangible mixed reality product design concept. Figure 4.17 is the design team's convergence to principles that will drive the tangible and digital design and related intended interactions. These thematic criteria and user-defined terms become significant guiding rods to the design craft. Form, color, aesthetics, communication, and brand are derived from these higher level themes.

Figures 4.6 through 4.17 are the compilation of design ideas based on the extensive longitudinal multi-method research approaches. Through the movement between Framing, Doing, and Making quadrants, the team is completely aligned with a purpose to Define the final design solution. The team has a repertoire of contextual knowledge and real-life scenarios to draw upon because of the rigorous, self-built, custom research process. Because of their holistic understanding of the

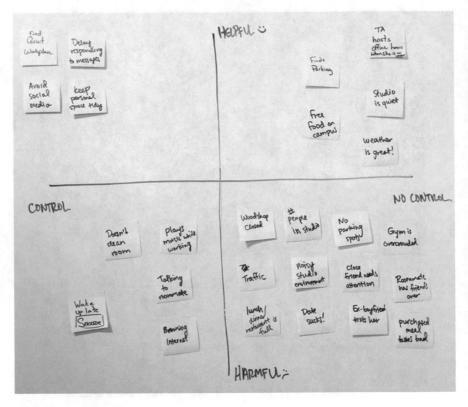

Fig. 4.9   Control and helpfulness

user group's needs and mindset, they are confident in converging on a single design that is a reflection of prior activities and output.

This team developed a video sketch demonstrating the future reality intended by the product system. Team Compass' video shows the extent of the new product interface and application of mixed augmented reality for enabling ADHD users. The true testament of this project is that the design concept they proposed could be transferred to the general population, alternate industries, and other environments to be universally applicable. https://vimeo.com/194847651

Team Compass: Katherine Apostolou, Elaine Choi, Gabriel Mitchell

This is relatively typical of the range and depth of research processes conducted in the UX and Product Design Studio courses. At the meta-level, the Compass team recognized that many ADHD students both need an external aid, but a customizable tool that fits and supports their own way of doing things. The transition for research

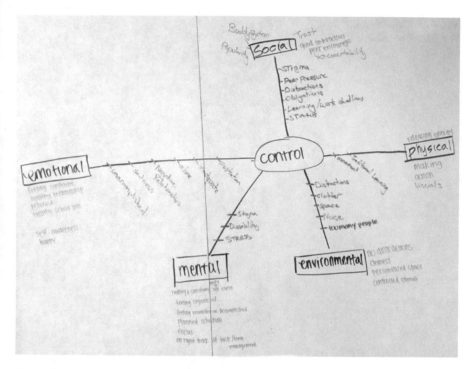

Fig. 4.10 Mapping user aspirations of control

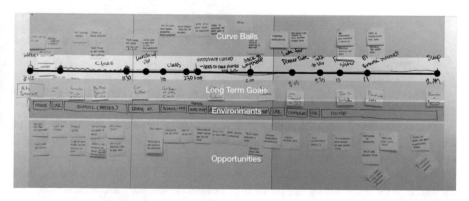

Fig. 4.11 Journey mapping

findings to conversion into the right side of the double diamond is expanded upon in Chap. 5. Even though Fig. 4.18 represents the final design solution, the value of the laborious research process allows for many other design solutions to be proposed. Various design iterations were completed in relation to the criteria determined by the preceding research processes that are completely viable in other

Fig. 4.12 Drawing it out
as dynamic scheduling

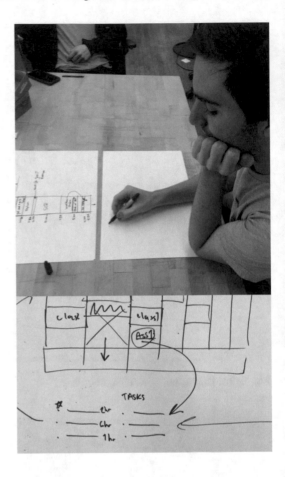

markets and applications. The value of a generative process can deliver beyond a singular solution.

This project reinforces the necessity to have a people-centered focus throughout each quadrant and design process. Though each team and project is highly unique in relation to the subject matter chosen, participants, and overall context—it is the tenet of placing people first, that forms the start and end principles of and for both Design at a meta-level and design of form, brand, aesthetics, and interaction. The purpose of this summarized design process is to portray key activities and output that are significant to the decision points in open-ended project briefs. These types of challenges are highly dynamic and require creativity, analytical abilities, methods for synthesizing what is relevant while taking relative risks to propose new and innovative ideas that are feasible for manufacture and market viable. Designers and engineers need to be agile in their ability to pivot and adjust depending on the situation. This takes time and experience. And the academic setting needs to provide these types of situations for experimenting, simulating, and ultimately delivering on complex open-ended design problems.

Fig. 4.13 Workshop: time and productivity

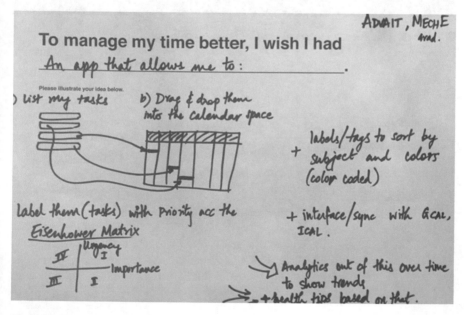

Fig. 4.14 User category themes, dynamic scheduling

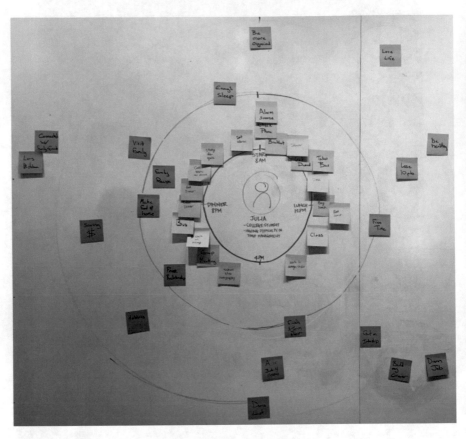

Fig. 4.15 Mapping of day-to-day tasks and longer-term goals.

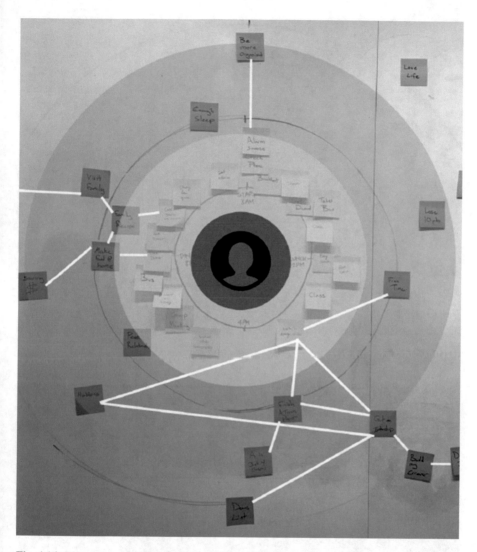

Fig. 4.16 Mapping tasks and goal relationships

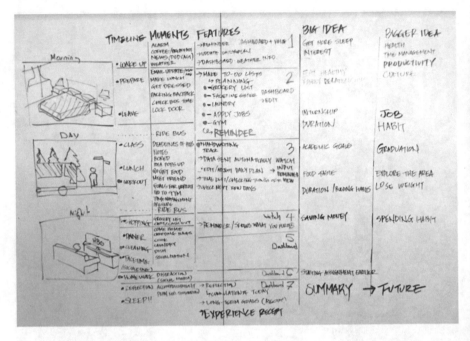

Fig. 4.17 Mapping of design concept timeline moment features

Fig. 4.18 AR mixed reality design concept video screenshot

References

Bloomberg Fortune. (2017). *Uber investors are selling to SoftBank at a steep valuation discount.* Retrieved February 17, 2018, from http://fortune.com/2017/12/28/uber-softbank-deal/.

Dubberly, H. (2017). Retrieved February 18, 2018, from http://www.dubberly.com/concept-maps/creating-concept-maps.html#more-1381.

Forbes. (2017). *#70 General Motors.* Retrieved February 17, 2018, from https://www.forbes.com/companies/general-motors/.

Gallagher, S., & Zahavi, D. (2012). *The phenomenological mind.* London: Routledge.

Menon, S., Sinha, A., & Sreekantan, B. V. (2014). *Interdisciplinary perspectives on consciousness and the self.* New York, NY: Springer.

Novak, J. D., & Cañas, A. J. (2017). Retrieved January, 2018, from https://cmap.ihmc.us/docs/theory-of-concept-maps.

Orbe, M. P. (2009). Phenomenology. In S. Littlejohn & K. Foss (Eds.), *Encyclopedia of communication theory.* London: SAGE Publications.

Chapter 5
Enabling D/design Through Curriculum, Collaboration, and Cognition

D/design in Practice with Others

The prior chapter described in various detail the activities and output necessary to frame and reframe a situation. It also emphasizes the power of representation. There is an unsaid necessity to have ideas born into the world so that others can judge and respond to the idea vector. Collaborative work can only be done when a common denominator is built. One of the strongest common denominators is the visual representation of the intended solution in various states. Early in the design process, this may be a concept maps and 2D sketches of the new idea. Midstream, the visual representation may be foam core models that allow basic interactions between the user and device. And when more tightly defined, a prototype is offered before the gatekeepers bless the tool steel prior to CNC and EDM. Working collaboratively requires a medium in which all fields can visually and cognitively understand one another's intention while allowing progressive changes and adjustments.

Few people possess, and seldom want to solely shoulder, the burden of project responsibilities that come with significant costs before sending it to a tool and die maker and mass manufacture. CNC operators used to and some still read two-dimensional print sheets from CAD models. Thankfully much of this workflow now resides digitally. However, before reaching this apex, multiple stages of review and decisions have to be made prior to the signoff. And this requires multiple 2D and 3D output so that inclusive team discussions can occur. The key point of this section is to enable a highly collaborative environment. This needs to be in both digital and physical spaces. Just as Chap. 2 emphasized the necessity for a studio or studio-like venue to allow experimentation and critique of all forms, it is this time and space platform that allows collaboration to exist and flourish. The ability to set the stage for individuals and teams to have critical discussions is essential. Essential in the sense that the old adage that the sum of the parts is greater than the whole.

One such instance occurred with a vendor bidding for the construction of the 8T MRI transfer table. Having had several on the phone meetings, the OSU engineering

© Springer Nature Switzerland AG 2019
W. C. Chung, *The Praxis of Product Design in Collaboration with Engineering*,
https://doi.org/10.1007/978-3-319-95501-8_5

and design team traveled several hundred miles to their offices. We brought the large orthographic drawings and unrolled them to talk about the functional requirements. Within minutes of discussion, the highly defined drawings were marked up and sketched upon as a fluid discussion transpired. There was a single half-second moment of the vendor's engineers pause to see the seemingly finished state of the clean orthographic plotted drawings, suddenly dashed away with freehand markings across the page. Once they understood that these seemingly clean and "correct" drawings were not precious and just part of the process to arrive at a better solution, their mindset and own actions changed. They too picked up a pen and started reimagining, reframing, and remaking the prior version into something better suited for our end solution. The common denominator of that day was the large 2D plotted orthographics. On another day it was a full-scale foam core model. But the essential elements included all of us in the same room at the same time, having the ability to rework the problem at hand through a visual medium. This type of higher level order of communication and decision-making only occurs when the visual and tangible assets are present. The higher the fidelity of the medium or models, the more narrow the discussions. There is a nuanced reason for not immediately fast forwarding to highly finished or polished designs if the team wants external input and feedback. However, as the reader can see, there are many elements at play. The combination of the right environment, activities, and output allow for effective team collaboration. And in this particular case, getting nine uniquely talented people in that room, allowed us to solve the design, engineering, and fabrication challenges.

But how can we teach young designers and engineers to work on a range of projects that result in innovative solutions? Innovation incubates best when integration and participation of all disciplines are active during key parts of the product development process. And interdisciplinary collaboration is not easy.

Eastman, compiled a comprehensive book, Design Knowing and Learning: Cognition in Design Education (2001) which brought perspectives from engineering and architecture educators. Esteemed academics studying student design activities through various methods reflect on addressing professional standards and in general how their disciplines use, apply, and perceive design. Critical questions such as whether the domain of design can be separate from their prescribed disciplines, how does the design discipline differ from their own, and is it even possible to study such a complex and all-encompassing term antiseptically. This dense compilation of discourse helps to discern between the implicit and explicit educational differences of each discipline. A form of this research was possible when working with colleagues at my current institution.

I had the fortune of co-teaching with Conrad Zapanta, Ph.D. at CMU's capstone biomedical engineering course over a 5-year period. Populated by 45–60 students per year, a range of engineering disciplines such as chemical, civil and environmental, electrical, materials science, and mechanical engineers took the project-based course. Embedding studio-trained designers into the engineering teams were deemed a benefit to all the teams from post-interviews and instructor reviews. This longitudinal, two-semester or full academic year of 9 months, was coupled with an external client or research institution to develop solutions for a variety of medical and health challenges. Various tools such as the Comprehensive Assessment of Team Member Effectiveness system (CATME) allowed for best practices of team creation, self-evaluations, and periodic peer evaluations. Teams with students from four or five different majors resulted in higher average scores across several categories compared to teams with only one or two majors. Over the past years, several teams have been recognized by winning design awards and external competitions. On two separate occasions, two teams proceeded through incubators, received seed funding, and developed a viable product that is on the market or being produced (Zapanta, Chung, & Dickert, 2017). To accomplish synergies and understanding among such diverse majors, these soft-skills need to be practiced and enacted on a regular basis.

The Force Multiplier Affect: Diversity in Teams

This chapter will only touch upon the unique characteristics of personality and organizational behavior and how it affects team dynamics. Though this is one of many components to setting up an innovative team, it is one of the most powerful force multipliers available to design or any field. Designers and engineers who have had both good and bad project experiences realize that this uncontrollable factor can make all the difference. The difference can be as mundane such as accomplishing tasks on time and within certain parameters. And there are times when another team member brings forth a different way of thinking and working that elevates the entire team's performance and expectations. Collaboration occurs when each person respects one another and each can see the value in either the proposed ideas or processes. Though possibly an oversimplified statement, the essential starting point requires the latitude and time to listen to one another. I have had several experiences as a professor observing engineering and design teams not give each other the benefit of the doubt. Both sides have a preconceived notion of each other's roles and have already relegated one another to the corresponding tasks. Unfortunately, Design is not just about individualistic tasks as we saw in the Design Praxis Matrix. Everyone can play a part in each quadrant. Everyone's experience and point of view can compound the quality of actions and output that comes from each area. And by applying these team actions and output, framing and reframing results in D/design perspectives and eventual innovative solutions.

One of the main goals of this book is to show and share each discipline's approach to teaching and how each goes about designing. And by doing this, understand each other's perspectives so that each can reconcile their own unique attributes to move forward and work together in harmony. How can we maximize each other's current skill sets and mindset? Does goal oriented versus process-oriented workflow enable better collaboration? And how can we move forward to teach collaborative practices in our own courses and curriculum?

The term, team implies working together effectively and efficiently to accomplish certain goals within a time period. However, the subtle difference between highly collaborative teams which seek new discoveries and challenge the norm are ones that embrace risk, experimentation, and play. To attain this level of trust among your teammates requires confidence and respect for each other's core skills and abilities, but also a mutual friendship that allows for openness and humor that is and is not always about the project problem at hand. It is relatively easy to spot teams that enjoy each other's company and others who are just going through the motions. Empathy is not meant to be an application method that turns on and off when convenient. Empathy is a way of life in which the fellowship of your teammates is a constant. More than a mindset, it is a natural course of living and working. Lucas, Ballay, and McManus book, Trillions (2012) is based on years of experience managing a highly successful design consulting firm, MAYA. The company's personnel makeup is listed as makers, strategists, thinkers, artists, and technologists. Not exactly titles or majors you can pursue in a higher-education format. Thought reading between the lines, they are computer scientists, anthropologists, visual designers, architects, and many more. Each has their own specialty and depth. However, Ballay speaks to the mindset necessary to allow all of these talented people to work together. He calls it "Violent Agreement and the Opposable Mind" in response to queries of how the principals can formulate such synergy and culture amongst different ilk of folk. Ballay quotes: F. Scott Fitzgerald stating "The test of a first-rate intelligence is the ability to hold two opposing ideas in the mind at the same time, and still retain the ability to function." Ballay further states "If the group acknowledges that interdisciplinary collaboration is just going to be hard, they almost always find their way to a deep, fundamental agreement that has been masked by different languages and descriptions. Once they get past that, their underlying concepts are compatible and often mutually enhancing. They find they can combine their ideas to make deeper, rounder, more enduring worldview. It's a phenomenon we've seen time and again, but it demands an investment." (Lucas et al., 2012)

The T-Shaped Designer and Other Letter Shapes

The reasons and purposes for a young person to pursue higher education is diverse. Some may enter the institution because it continues the path of intellectual and personal growth. Others expect to be taught immediately applicable job-specific skills so they are marketable and productive in the workforce. Somewhere in between is a hybrid of these two major thought camps. However, without exposure to other disciplines and majors, a future employee may not be able to grasp or understand their and other team members' roles when it comes to complex projects. Tim Brown, CEO of the well renowned consulting firm IDEO coined the term T-shaped designer. This image was to help describe people who possess a breadth of general knowledge (the horizontal top) and a depth of skills and experiences (vertical descender stroke) (Fig. 5.1). Brown recognized that company success was directly indicative of the types of people working together. And the type of people who did well in teams were those who had empathy for other disciplines. It should be noted that IDEO is made up of a broad range of disciplines—people with depth in

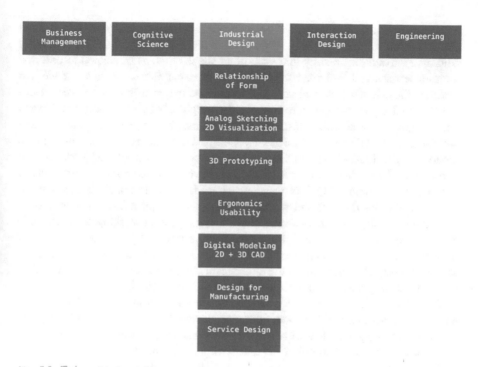

Fig. 5.1 T-shaped industrial designer

anthropology, design, engineering, medicine, and many other humanities and sciences. Brown's interview quote from Chief Executive magazine best sums it up (Hansen, 2010):

The vertical stroke of the "T" is a depth of skill that allows them to contribute to the creative process. That can be from any number of different fields: an industrial designer, an architect, a social scientist, a business specialist, or a mechanical engineer. The horizontal stroke of the "T" is the disposition for collaboration across disciplines. It is composed of two things. First, empathy. It is important because it allows people to imagine the problem from another perspective—to stand in somebody else's shoes. Second, they tend to get very enthusiastic about other people's disciplines, to the point that they may actually start to practice them. T-shaped people have both depth and breadth in their skills.

Figure 5.1 is specific to what a traditional industrial designer T-shape knowledge and skills may possess. Having some foundation and exposure to engineering, design for manufacture and assembly, and other technical areas in both the horizontal and vertical axes allow the designer to be aware of the importance of these core principles. These aspects in turn allow the designer to better understand the driving factors for engineering logic, decisions, and concerns. And in theory and practice the collaboration is more synergistic if the engineer's T-shaped attributes have some design exposure and experience.

Brown contrasts this to an "I" shaped unidimensional designer with limited ability to connect to other seemingly disparate disciplines. There are other authors who have carried the analogy afar with other shapes such as M-shaped to describe multi-expertises and E-shaped for various work experiences. However, if we just maintain the original T-shaped analogy, what is the major difference between a studio-trained designer compared to a mechanical engineer? Does the horizontal broad knowledge for engineering consist of mathematics, physics, thermodynamics, and other science-based courses? In Brown's definition of enabling empathy through the broad horizontal knowledge, this model may not fit engineering's early educational curriculum. Rather the T-shape is most likely inverted when considering a classic engineering education. Much of the coursework is about deep and specific knowledge before branching out and working with other disciplines. Albeit that knowledge is a necessity for understanding, building, and testing working structures and designs. However, a balance of both specification oriented goals and open-ended problems early in the education may benefit a student to have a more flexible mindset. Otherwise, when they are in the upper-level classes and eventually in the workforce, connecting and collaborating will be an increased challenge.

One of the main purposes of a holistic education is to provide a platform that enables students to hone their own competencies while benefiting from their peers who are attempting to do the same. As simplistic as this may sound, it is very similar to a sports team going through drills and scrimmages. The studio is a shared space with your current cohort as your support team. However, there is a fine line between collaborative competitiveness versus blatant competitiveness. Those who work in a studio environment solely for themselves will not benefit from the shared skill building, honest critique and reflections, and camaraderie that continues into the professional environment. Thus collaborating and learning to work together, should start immediately in the higher-educational environment.

Course Content and Timing Matters

The School of Design's Product Design track philosophy is based on the interaction between person and object. This mantra is taught early and constantly through the curriculum. The course structure is shown in Figs. 5.2 and 5.3, which show that once students have beginning skills in Form and Context, their Year Two, second semester is explicitly Design for Interactions. Each design studio in the first and second year is coupled with a design lab. The studio lab concentrates on the technical skills, machines, and software required for the trade. The design studio sets the stage for what and how to approach a problem. The paired studio and lab courses work in tandem to build off each other's content. Analogous to the Making and Doing quadrant, the activities and output mirror the intent of learning by actively experimenting and reflecting. In contrast to other visual and performing art fields such as a music conservatory method, atelier system, or an apprentice model, the design studio + lab model is a mix of teacher and peers working and learning

BDes: Year One

1 **Fall** Freshman

Explore basic concepts in design by deconstructing products, communications, and environments. Examine interactions in the built and natural worlds from a design perspective.

Studio: **Survey of Design** 12 units

Investigate how design works in the world, and learn to describe your thinking about design through sketching, modeling, photography, and video.

Visualizing 9 units

Use visual design tools and methods to communicate ideas.

Placing 9 units

Examine identity, values, and worldview through the history of the local bio-region.

Computing @ Carnegie Mellon 3 units

English 9 units

Psychology 9 units

2 **Spring** Freshman

Work on projects to develop skills across all three design tracks: products, communications, and environments. Enhance your understanding of the broader impact of design in the world.

Design Lab & Workshop 12 units

Practice design approaches, methods, and tools in focused 5-week sessions.

Collaborative Visualizing 9 units

Learn to work with others to visually generate, iterate, and refine ideas.

Systems 9 units

Draw models of relationships within systems to identify where and how to intervene.

Photo Design 9 units

Learn how to deconstruct and construct images in the world and in the studio.

Global History 9 units

Fig. 5.2 CMU SOD BDes year 1

together, though the one-to-one design student-to-teacher ratio is unsustainable from several points of view. However, to carry the music analogy forward, the design studio is more similar to an orchestra or band constantly working together. The individual practicing scales and shifting is the equivalent of sketching and

BDes: Year Two

3 **Fall** Sophomore

Customize your pathway by focusing on two design tracks of your choice. Investigate the roles that physical, visual, and digital forms play in our lives, and learn methods to understand how people think and work.

Studio I: Form & Context   9 units

Practice design approaches and methods in two of the three design tracks in focused 7.5-week sessions.

Prototyping Lab   9 units

Refine skills using design tools in two of the three design tracks in focused 7.5-week sessions.

How People Work   9 units

Study basic principles of human-centered design and research methods.

Design Elective   9 units

Build skills like drawing, programming, learning software, imaging, etc.

Academic Elective   9 units

4 **Spring** Sophomore

Take a deeper dive into a single design track. Build knowledge and hone skills in the design discipline that interests you most, and learn methods and processes that help you develop and refine your ideas.

Studio II: Design for Interactions   9 units

Focus on one of the tracks you studied in the third semester.

Prototyping Workshop   9 units

Develop skills to prototype and iterate design concepts.

Research Methods   4.5 units

Learn about exploratory and ethnographic methods; participatory and generative design; evaluation and testing.

Cultures   4.5 units

Understand differences between people stemming from ethnicity, gender, class, etc.

Academic Elective   9 units

Free Elective   9 units

Fig. 5.3 CMU SOD BDes year 2

modeling. Whereas when the individual musician is amongst a group of equiva-lently practiced and accomplished players, the combination is unachievable alone.

The engineering education model of lecture and lab is similar in intent. The main concepts to be learned are traditionally presented in a large lecture hall with exam-ples and case studies. At a later time and date, students attend smaller class lab. If there is no hands-on work, a recitation would be the equivalent. A subtle but signifi-cant difference in engineering recitation and lab is the assigned instructor. In most engineering cases, a Teaching Assistant (TA) or Graduate Assistant (GA) is the primary instructor. The quality and experience may vary significantly from year to year and class to class. Whereas product design studios are almost always staffed by a full-time professor, not a graduate student. If there is a TA or GA, they are in sup-port of the design professor and act as a secondary voice. Just as all assistants, they can vary in both teaching and professional experience. Thankfully, the design studio structure has the professor as the primary instructor. This structure is predicated on a larger mechanism though. One of these includes the predominant number and scale of each respective discipline. At the time of this print, Carnegie Mellon University's College of Engineering reports 1780 full-time undergraduates, 1850 graduate students, and 184 faculty members. The School of Design student numbers were 180 undergraduates, 50 graduates, and 33 faculty members. Also engineering sponsors and adds a significant number of doctorate and PhD students to their fac-ulty. Some are dedicated to research and may not see the inside of a classroom. But the structure of an academic unit is formed to deliver courses to an x number of students every year. And these Masters and PhD students factor into the teaching quota and delivery. Though the design field and CMU School of Design do have PhD and DDes degrees offered, the number of these students is less than a dozen. And typically half will teach into a lecture (non-studio) design course or elective courses. This administrative structure does have an effect on the manner and type of educational delivery. The explicit difference between theory and practice becomes quickly self-evident. A graduate student or any instructor who has limited external professional practice may not have the ability or experiences to transition from the book material, theory to real-world praxis. The structure of the coursework allows for certain efficiencies to be achieved in the engineering school. Correct answers and optimal answers are clear-cut. However, it is more difficult to staff and manage highly complex and open-ended coursework.

Providing Platforms for Individual Learning Experiences

To shift from the organizational structure details that are built to serve a larger mechanism, let us consider how humans think and learn. Though the design stu-dio curriculum was most likely not conceived with explicit analysis of how the human brain works, using Rumelhart and Norman's (1976) simplified categories of cognition and learning, highlights the benefits of the studio-based education. Norman, a renowned cognitive scientist describes two types of cognition and three types of learning (Norman, 1994). It is acknowledged that the brain is much more complex than five different categories. But the value in this is to highlight the major activities and differences between each. Norman considers two types of cognition: experiential and reflective. "Experiential thought is reactive, automatic thought, driven by patterns of information arriving at our senses, but dependent upon a large reservoir of experience." Norman uses examples of a pilot acting and reacting instinctively to the state and position of the airplane, a musician playing a piece of music effortlessly, or simply driving home from work. Reflective thought is "that of concepts, of planning, and reconsideration." It requires dedicated time and deliberate energy aimed at the situation. If we take the pilot example, reflective thought is necessary to plot the course to a destination. When conditions change such as bad weather in the original flight path, a pilot must assess the state of the plane, how much fuel is onboard in relation to an alternative route, and make several decisions. Moreover, reflective thought can be slow and laborious. This type of thinking "requires the ability to store temporary results, to make inferences from stored knowledge, and to follow chains of reasoning backward and forward."

Norman describes the three type of learning as accretion, tuning, and restructuring. Accretion is described as learning facts and daily accumulation of information. The acquisition of memories of prior events adds to a person's knowledge base incrementally. Per Norman and Rumelhart, accretion is the normal learning that has been most studied by psychologists—learning of lists, dates, names of presidents, phone numbers, and related things. This type of learning presumably occurs through exposure to the concepts, a processing stage that transforms the information into a memory representation and it added to a person's database of knowledge. At this stage, no structural changes in the information-processing take place and it is considered easy, painless, and efficient. The second type of learning called, tuning can be best described as practicing. As a person performs a new skill, a considerable amount of energy and concentration is required. This may apply to motor skills in a sport or cognitive tasks such as multiplication tables. Tuning can be a very slow and laborious process where the mind and body are syncing. Playing a musical instrument while learning a new piece of music is a prime example. After several hours and attempts, does a piece become memorized and the muscle memory follows the correct notes and tempo. Alternately, when we do not practice or engage in tuning, that skill typically atrophies and worsens. Lastly, restructuring is considered forming the right conceptual model. If accretion and tuning are primarily experiential modes, then restructuring is reflective. Restructuring requires processing of previous facts and knowledge that changes in the categories or schemata. "Through tuning an existing schema is modified. Through restructuring new schemata are created." (Rumelhart & Norman, 1976) The hard part for Norman was improving upon the rate of learning. And like most things, if the person was unmotivated to accomplish a task or goal, the efficacy was low.

The segue into this simplified but fundamental way of human thinking and behavior is to reinforce the ways and means of helping students maximize their potential. Chapter 2 covers how the right environments support a range of learning by through access, learning by doing, critique, and mimicking through practicing. This chapter shows the inner workings of using that space and collaborative processing to elevate a studio student's thinking, making, and doing. And to be creative and innovative requires mental effort that moves beyond regurgitation of known facts and givens. The peer co-location and informal and formal critique, support reflection of self and others' work become a necessity to learning. This recognizes that learning and teaching benefits from both solo practice and constant classmate and team interactions. These interactions should be between other peer disciplines so that cross-disciplinary domain knowledge can be integrated into the project process and eventual solution. Design studio education is a generative learning process that requires continual cultivation throughout the undergraduate experience.

Flip this concept, and these can be new skills, approaches, thought processes, or modifications to existing practices. One particular student may be the more advanced peer teaching another student. An old educational anecdote states, there is no better way of learning, than by teaching that subject. Thus the studio educated student implicitly has a responsibility to both themselves and others. These types of peer-to-peer and formal teacher studio class experiences are essential for project-based curriculums. D/design is a confluence of multilevel interactions with other people and disciplines. Setting the learning and practicing stage for these complex elements and activities to take place is the challenge for educators and institutions. Recognizing what tools, space, and timing allows a person to thrive and learn is essential to the future and culture of innovation and invention.

The following case study is a traditional collaborative project with classic design and engineering roles. The story portrays differences between how designers and engineers work, and in turn, shows the disconnect and inability to be flexible and in thinking and working together. This case was intentionally placed later within the series of book chapters to enable the reader to see the many positive aspects of multidisciplinary collaborations, first. It this case, there are breakdowns in communication that stemmed from ill-defined roles, nebulous and fuzzy front end problem framing, and lack of experience between team composition that creates unwanted tension and conflict.

However, once a clear set of goals and sub-goals were established, appropriate paths and processes to empower each team member fell into place. In retrospect, each team member's role was essential to competing this complex and challenging project. We needed each engineer to call on their deep knowledge and tools, just as the designers did the same. And to note, after completion of this relatively technically difficult project, the engineering and design team's positive synergy carried them through a subsequent large animal veterinary CT scanner transfer bed project and continued life-long friendships.

Case Study: 8T MRI Design of Transfer Table and Gantry Design

The principal investigator, Dr. Pierre-Marie Robitaille assembled a team to develop and design a fully functioning clinical 8T MRI system at The Ohio State University's Medical School. To put things into perspective, the power of this research magnet was the most powerful clinical device proposed in the world just before the year 2000. Typical commercial MRIs in current hospitals and institutions are 1.5T. Though there are many MRIs that are more powerful, they are very small coils and enclosures. None had attempted to build a bore size of 80 cm in diameter or capable of scanning a human body at this high Tesla power. The copper windings of the main magnet spanned several hundred kilometers. The assembled team consisted of three industrial designers, two mechanical engineers, and two electrical engineers. However, the initial engineering and design team was wrought with team interaction problems. The first design team consisted of four undergraduate mechanical engineering students and three graduate product design students. Much of the background research and definition of the design problem was formed by this first group. They also were able to accomplish the Kevlar and fiberglass patient table mock-up and scale models of the mobile patient table, gantry, and MRI room environment. However, friction among the student mechanical engineering and product design teams existed because of several reasons. Primarily, the project required trust and patience in each other's abilities. The product design discipline relies on mock-ups and lower fidelity modeling to decide on final designs. In the case of the engineering discipline, their decisions sometimes require extended time periods of near-finished prototypes and various modeling analysis before fabricating the final model. These differences in work style, timing, and use of different modeling mediums for different purposes invoked team conflict. Expectations of the engineering cohort having practical experience in making and fabricating, and a design process to rely upon, was unrealistic. In addition, some of the students' inability to grasp the entire system rather than hyper-focusing on a component, hindered the progress of the team. The engineering members jumped directly from background research to working in the far right diamond.

Thankfully, Dr. Robert E. Bailey Mechanical and Nuclear Engineering professor reorganized the team with one mechanical engineering graduate student, one electrical engineering graduate student, two product design undergraduate students, and three faculty members in each of the respective disciplines. All of the persons on the team had some form of professional and practical experience. Most importantly, each of the team members had worked in an interdisciplinary environment prior to this experience. The ability to trust and respect each other's ability, contributions, and work flow was the crux of establishing an effective problem-solving team.

MRI System Engineering Requirements

The following paragraphs describe the multiple layers of decisions necessary to meet the specific functional requirements of the MRI patient transfer table and gantry system. Fundamental criteria included: accurate control position of a 275 cm (9 ft) long cantilever patient table, highly accurate patient table horizontal and vertical movement and positioning, address patient comfort for prolonged scanning, and meeting primary medical users (technicians, researchers, and doctors) functional needs. Mechanical engineering's first activities and output was CAD modeling and running various simulations and analysis.

The final design for the 8T MRI gantry and the mobile patient table was dictated by the highly specific and functional criteria. The list of product requirements included the accurate movement of a 275 cm cantilever table and low deflection at the magnet's isocenter. In addition to this requirement, no pressure or weight could be transferred to the magnet bore walls. The back-end of the magnet bore was covered, eliminating the option to run supports through the bore. The patient sled was designed to carry a point-load of 46 kg (100 lb) at the very end of the cantilever transfer table with <10 mm deflection. The material selection for the gantry, MPT, and other materials and components were also quite limited by the inherent high magnetic field and resulting attraction force effect hazard. Non-ferrous materials, such as aluminum, austenitic stainless steel, and polymers were selected to take into account these unique and dangerous conditions. The system required a mobile, cantilevered table with patient sled drive and height control mechanisms, and a gantry which would house electronic controls and displays connecting the computer systems to the patient table when docked. The mobile patient table was designed with side fold-up trays for administering contrast or intravenous medications to the patient prior to being scanned. The internal shape of the sled also allowed air ventilation into the bore. Because of the unique size and dimensions of this specific magnet, utilizing prior designs of 1.5 commercially available transfer tables was not viable.

Material selection needed to be chosen and tested carefully. The most important known dangers are the projectile effect, which involves the forceful attraction of objects with magnetic properties, unnecessary absorption of RF power, and bioelectric effects of moving a patient too rapidly near the 8T magnetic field. With the super high magnetic field, the design constraints require the avoidance of using any ferrous materials, including certain grades of stainless steel at particular distances. Secondly, the magnetic high frequency and eddy currents prohibited electrically conductive materials, such as carbon fiber in close proximity to the MRI magnet–gantry interface. Even the use of non-ferrous metals had to be strategically placed in relation to the perpendicular Foucault currents. The team opted for fiberglass as the main material for the entire gantry and external housing to cover the MPT mechanisms. The fiberglass thickness and layers ranged from 3.175 to 6.350 mm (1/8″–1/4″) were deemed adequate for repeated assembly and disassembly. The ability to drill and add, epoxy secondary parts, allowed the team with additional options and last minute customization. Fiberglass A-surface finish made it the material of choice.

The structure and chassis of the cantilever transfer table were primarily made of 6061 aluminum, 7075 aluminum, and austenitic stainless steel. A two-step drawer slide, much like a file cabinet drawer slide mechanism, was employed to allow the table to span the unusually long distance to the MRI isocenter. A short aluminum slide telescopic support would extend from the main frame (50.8 cm, 20 in.), then the full length of the mobile patient table would reach the isocenter 236.22 cm (93"). Polymer phenolics were custom milled to connect or cap pieces of the aluminum frame's open sections.

The Kevlar and fiberglass composite patient sled were manufactured by Hadlock Plastics, Geneva, OH. The viability of Kevlar, E-glass, and S-glass, combinations were prototyped prior to the final table design. The basic mechanical issues were concerned with the compression and tensile forces and the stiffness of the table at the extended distances necessary to reach the isocenter of the bore.

The docking mechanism of the mobile patient table consists of a support located at the magnet face which extends through the floor of the room and is anchored to the steel structure, which surrounds the 8T magnet room. Furthermore, this very rigid support contains connections to supply air and electricity to the mobile patient table.

The casters were purchased from Colson Caster Corporation and are rated at 350 lb. each. However, due to the high content of the steel, particular parts of the caster (except the bearings) were replaced with 7075 aluminum. The main stem of the caster was redesigned to make it a more stable attachment to the MPT's base frame. To allow low rolling resistance with no electromagnetic properties, ceramic bearings made by Sapphire Engineering Inc. from Tetragonal Zirconia Polycrystals (TZP)—SEI 1000 were installed within the patient transfer table. These particular bearings were rated as having a hardness of 1300 kg/mm^2 (Vickers) and a Young Modulus of 207 GPa. Fasteners were limited to nylon, austenitic stainless steel, or brass.

This relatively short description is necessary to provide a glimpse into the highly complex decisions required to design and build a working medical product system. This is in no way exhaustive of the incredible amount of time, detail, costs, and energy necessary for all the team members and external manufacturing vendors. But this illustrates the extremely narrow design and engineering options available to the team to accomplish the technical requirements necessary to achieve the MRI system's goals: the goals of moving a patient or test phantom from a prep room into the MRI room; docking and registering the patient table to the floor for positioning and AC power; and the ability to digitally control the patient table from the gantry.

MRI Product Design Patient-First Approach

In contrast to engineering's immediate activities to CAD model every part, develop test table material experiments, and determine constants and variables, design's first instinct is to talk to people. Identify who are the stakeholders and what kind of interview or research protocol can be employed to learn as much as possible from

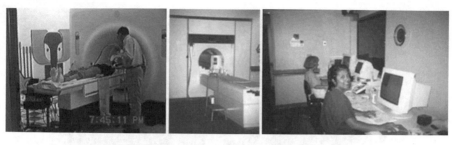

Fig. 5.4 MRI stakeholder research

Fig. 5.5 Medical surplus research

first-person research. How can we learn from others who have experienced a similar challenge? And how can we use these stories to help guide our design?

The design philosophy for product design is steeped in user-centered and research-based principles. The initial steps to understanding the problem and overall system reside with the actual stakeholders and interaction between the product and them in situ. Background research of current products was conducted through literature searches. On-site interviews with physicists, technicians, nurses, and radiologists at three MRI facilities were conducted (Fig. 5.4). This research provided an understanding of three distinct MRI medical systems, patient and technician interactions, patient to technician and equipment interactions, and the clinical scenario the stakeholders experience to obtain MRI images. From outside of the MRI room, video ethnography was used to capture the sequence of the medical technicians and patient interactions, the motion and time needed for the procedure, and the verbal communication. Eighteen patients were observed going through an MRI procedure. In addition, each team experienced the MRI procedure at least once. The design team was ignorant of the clinical hospital procedure that technicians are wholly responsible for the scans.

Interviews with an MRI parts manufacturer Hadlock Plastics provided the design and mechanical engineering team with insights into production techniques and material selections. An on-site tour of a fiberglass manufacturer, Skid-do Industries provided manufacturing techniques for the exterior panel and gantry designs. Lastly, an on-site visit with CTR Surplus, an MRI and medical equipment refurbishing and recycling company, aided in comprehension of prior MRI equipment designs (Fig. 5.5). Decommissioned MRI, CT scanners, and other large-scale nuclear and X-ray equipment were disassembled and available for tear-down observation and

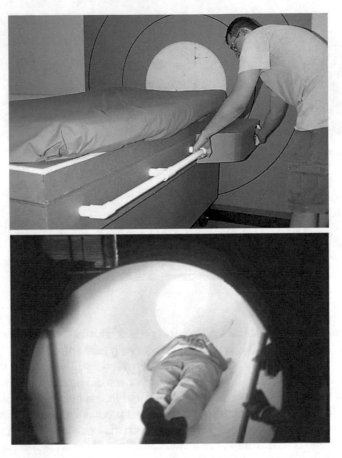

Fig. 5.6 Mixed fidelity modeling

analysis. Overall, all of this background research was necessary for the team to begin understanding the clinical and medical environments, interactions, and experiences. Immersion into a problem space is the first stage of a product and engineer's process. Through first-person accounts, developing their own research protocols to contextually understand the nuances of everyday experiences, become the essence of understanding and empathy for a product designer.

The project entailed the gamut of physical and virtual prototyping before fabrication. The typical design process of 2D concept sketches, 3D scale mock-ups, full-scale gantry mockups, and computer models was made. The extensive use of both physical and digital modeling allowed the client and engineering and design team to determine fabrication decisions and changes throughout the process. The combination of low-fidelity full-scale modeling with foam core and cardboard allowed for human to control surface simulations. Figure 5.6 used the low-fidelity material modeling with the actual specified bore lighting fiber optics track. High-fidelity physical scale models and 3D digital models were made to show the expected final

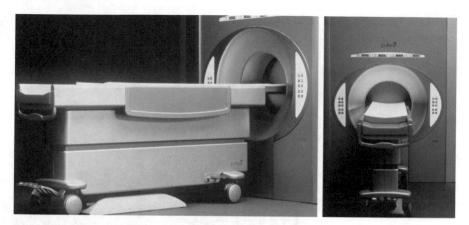

Fig. 5.7 High fidelity scale modeling. Valenciano, M. Youger, J

designs through the range of visualization methods. See Figs. 5.7 and 5.8 for examples. One unique video animation technique of green-screen compositing was used because of the size of the final product. This standard technology application in television and movie industries, allowed the client and team to visualize the actual human scale and intended interactions in relation to the large medical system. Figure 5.9 is a screenshot of green-screen composite.

For a project like this, the various models and mediums were needed to effectively communicate the design intent between the team members and the client. Each model helps the team to internally refine and understand the final output. Without the range of affordances of both physical mock-ups and virtual simulation, the design decision-making process would not have been as well informed or executed. Studio-trained designers and engineers will naturally use a range of modeling mediums to understand the human to machine interactions. And in turn, their subsequent design decisions match the intended tasks and the overall goal of the man–machine system (Fig. 5.10).

Fig. 5.8 High fidelity 3D
digital modeling,
Valenciano, M

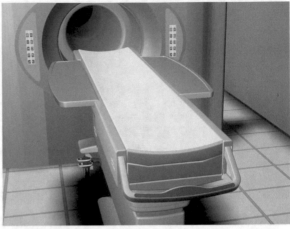

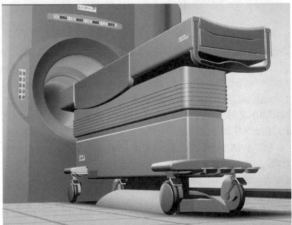

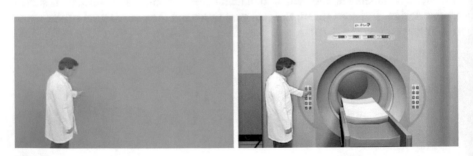

Fig. 5.9 Green-screen compositing, Valenciano, M

Fig. 5.10 8T MRI gantry and mobile transfer table, M. L., Jagadeesh, J., Bailey, R. E., Chung, W., Somawiharja, Y., Valenciano, M., & Youger, J

References

Eastman, C. (2001), *Design knowing and learning. Cognition in design education* (1st ed.). Amsterdam: Elsevier Science.

Hansen, M. (2010). T-Shaped Stars: The Backbone of IDEO's Collaborative Culture. *Chief Executive online Magazine.* Retrieved February, 2018, from https://chiefexecutive.net/ideo-ceo-tim-brown-t-shaped-stars-the-backbone-of-ideoaes-collaborative-culture__trashed/.

Lucas, P., Ballay, J., & McManus, M. (2012). *Trillions: Thriving in the emerging information ecology.* New York, NY: Wiley.

Norman, D. (1994). *Things that make us smart: Defending human attributes in the age of the machine. William Patrick book.* Boston, MA: Addison-Wesley Longman Publishing Co., Inc..

Robitaille, P., Warner, R., Jagadeesh, J., Abduljalil, A., Kangarlu, A., Burgess, R., et al. (1999). Design and assembly of an 8 Tesla whole-body MR scanner. *Journal of Computer Assisted Tomography, 23*(6). Allen D. Elster Editor, Lippincott Williams & Wilkins.

Rumelhart, D. E., & Norman, D. A. (1976). *Accretion, tuning, and restructuring: Three modes of learning.*

Zapanta, C., Chung, W., & Dickert, J. (2017). Summer 2017 CURQ web vignette: Interdisciplinary design Teams for biomedical engineering design. *Council on Undergraduate Research Quarterly, 37*(4), 12–13. https://doi.org/10.18833/curq/37/4/17.

Chapter 6
Domain-Specific Tradecraft

Domain-Specific Tradecraft

The practice of a discipline's tradecraft typically requires domain-specific knowledge, the culmination of theoretical information coupled with hands-on experience, and immersive and prolonged application of duties related to the profession. The difference between any discipline can be distinguished by the person's unique ability to perform tasks or operations others cannot without specific education and training. Even though most people can play golf, they do not have the ability to step into a professional tournament and perform with others who have had extensive practice time, experience, and coaching. People self-qualify their skill level by saying "I golf" versus "my handicap is −3." The highly domain-specific learning and teaching modes of accretion, tuning, and restructuring makes the clear difference between a casual golfer, an experienced golfer, and a PGA professional. And even though a profession may have similar tools such as an artist, designer, engineer, inventor, or a makerspace patron, there is a distinct difference between each. The respective tradecraft separates an artist from a designer and from an engineer. Additionally, the mindset, intent, process, and end output are very distinct. Each profession's tradecraft separates the subtle and not so subtle differences even though they may be using the same instruments. A chef and a baker are very different in how they approach a dish, navigate a kitchen, measure ingredients, the tempo and speed of their processes, and end output.

Chapters 2 and 5 discuss the tangible assets that need to be in place to enable a beneficial studio learning experience. They also present some differences between the design and engineering curriculum and courses. However, if we look at some of the first historical establishments of mechanical engineering programs, their descriptions sound more similar to today's current design studio setting. There are many renowned programs, but these following anecdotes describe the environments and the establishment of a mechanical engineering approach to teaching. This excerpt

© Springer Nature Switzerland AG 2019
W. C. Chung, *The Praxis of Product Design in Collaboration with Engineering*,
https://doi.org/10.1007/978-3-319-95501-8_6

from University Michigan's Engineering history starts us with a significant military figure who devised how to engage and educate young student engineers. And the second short paragraph describes Georgia Institute of Technologies engineering's inception.

It was not until Mortimer E. Cooley, a naval officer, came to Ann Arbor in 1881 that ME at Michigan gained an independent identity. Cooley, an 1878 graduate of the US Naval Academy, was one of a number of naval officers appointed by Congress to university facilities. His assignment was to establish an ME program. Over the next three decades, his leadership laid the foundation for a thriving department.

In the beginning, the ME curriculum consisted of Workshop Appliances and Processes; Pattern Making, Moulding and Founding; Mechanical Laboratory Work (Shop Practice in Forging); Machinery and Prime Movers (Water Wheels and Steam Engines); Machine Design; Thermodynamics; Original Design; Estimates, Specifications, and Contracts; and Naval Architecture.

The first mechanical laboratory was built under Cooley in 1882. At the time, engineering classes were held in the South Wing of University Hall, but there was no laboratory building. To remedy the situation, Cooley used an appropriation of $2,500 from the Michigan legislature to construct and equip a two-story laboratory building. Plans for the building were put together by ME faculty member J. B. Davis. According to an earlier history, "it was a two-story structure of frame construction with bricks placed edgewise between the studding. The ground floor was divided into two rooms, the foundry on the east end and the forge shop, brass furnace, and engine room on the west. The foundry also included two flasks, other necessary foundry tools, and molding sand. It is important to note that the shop contained the first steam equipment in the ME Laboratory, a forge, anvil, tools, a brass furnace, and a four-horsepower vertical fire-box and steam engine. The second floor was also divided into two rooms, one of which was occupied by the pattern shop and the other by the machine shop. The equipment in these rooms consisted of a wood-turning lathe built by Cooley and members of his class, and an iron lathe, salvaged from the basement of University Hall and repaired by the students" (University of Michigan, n.d.).

https://me.engin.umich.edu/about/history

A school of technology was established in Atlanta in 1885. In October 1888 the Georgia School of Technology opened its doors and admitted its first engineering class: 129 mechanical engineering students enrolled in Tech's first degree program. As part of their education these early students worked at trades such as forging, woodworking, machining, and mechanical drawing (Georgia Institute of Technology, n.d.).

http://www.me.gatech.edu/about/history

Granted these more than a century old historical beginnings are the stepping stones to current day engineering programs. However, the described learning environments are about applied, hands-on experiences. And we still live in a world that requires moving matter and transferring energy. The ability to imagine, analyze, test, and assemble through appropriate material selection, mechanisms, and energy sources are some of the basic underpinnings of everyday life and current day mass manufacturing. Teaching these concepts has been altered with the use of digital technology and other teaching resources. So current day education may still have some form of exercises and tests in physical or digital mediums. However, undoubtedly every area of study has advanced digital methods and tools for modeling and simulations. And from these digital models, physical prototypes or even short-run production parts can be printed or manufactured. Significant benefits are achieved through these digital methods. And so the educational philosophies and decisions are very much about how and when to use these technologies to best prepare designers and engineers for professional careers. The professional world can demand highly specified models for production, devising devices for testing and evaluation, innovative thinking and building of solutions specific to a particular industry or task, and an ever-expanding list. So by referring back to the Defining quadrant of the Design Matrix, an important corollary is how each discipline defines their designed forms and ideas.

Chapter 3 assumes the final output will be used in the same way by other disciplines. However, this is far from the truth. There is value in highly specific, as close to final production as possible digital and physical modeling. However, at what cost and time do this limit, and hamper the design process? Most experienced designers and engineers have had an experience where the final design misses the mark. No matter how clean and polished the model, the left side of the double diamond may not have been adequately addressed. The value of low-fidelity modeling hedges between full commitment to a final design by garnering input from both internal team members and external suppliers, vendors, tooling, and the client. Some of the greatest value that comes from physical prototyping is the immediate ability to observe and discuss form, interaction, and the relationship between object and user in real time. Even if the design is an electronic or powered device, role-playing and imagining the scenario allows many assumptions and understandings to be addressed.

And with many academic subjects, the profession is always advancing and changing. The question is not about what can be produced, but what can we learn by producing them with appropriate resolution, speed, precision, and multitude. Thus education shifts in varied measures to meet these new changes. And the role of educating a young designer and engineer is still about enhancing their learning of a subject while providing them opportunities for investigation and discovery. Hopefully, the reader accepts that Design is more than just packaging, aesthetics, ergonomics, and material choices. Design is about challenging the norm with and through cognitive and creative skill sets. By generating new physical and digital objects, the interaction and discussion that occurs once the artifact is born into the world allows for the re-tuning, reflection, and reinterpretation of the current and future world between self and others. Figure 6.1 is relatively typical examples of the visual mediums used to build out the designed object. Each level of resolution inherently possesses a certain limit to the fidelity of form and function and eventually a commitment to material choice and subsequent manufacturing and assembly options (Figs. 6.2, 6.3, and 6.4).

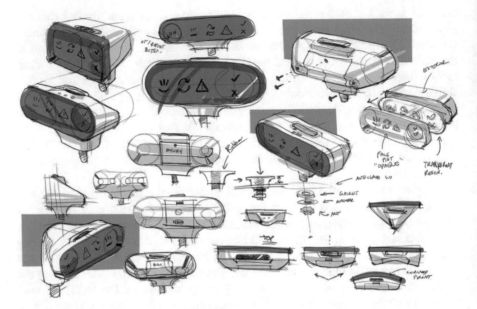

Fig. 6.1 Autoclave temperature sensor design visual medium example sketches, Kan, K. (2014). Carnegie Mellon University

Fig. 6.2 Autoclave temperature sensor design visual medium example low fidelity cardstock, Kan, K. (2014). Carnegie Mellon University

Setting the Stage for Self-Investigation and Discovery: Open-Ended Project Sets

Depending on whether an undergraduate studio program degree is a 3-year, 4-year, or 5-year program, the open-ended projects should be introduced after some proficiency in basic design skills (2D and 3D visual manipulation) has been established. Intermediate and advanced design briefs are open-ended, project-based challenges rather than defined problem sets which have narrow or predictable results. Though the curriculum structure was previously covered in the prior chapter, it is the project brief and resulting experiences within those courses that make the difference in

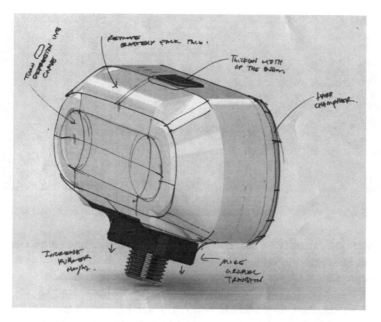

Fig. 6.3 Autoclave temperature sensor design visual medium example solidworks model printout markup, Kan, K. (2014). Carnegie Mellon University

Fig. 6.4 Autoclave temperature sensor design visual medium example solidworks model rendering, Kan, K. (2014). Carnegie Mellon University

creating an effective studio based and collaborative education. Recognizing this fact requires a detailed explanation of what makes a good open-ended design brief and challenge. The challenge level should allow a student designer to be self-motivated in the subject matter, set their own bar for areas that need growth, and means for peer and self-reflection and critique. One example course design brief that this author has established as a standard advanced level studio course brief is called the Low Tech, High Touch and High Tech, High Touch project:

You will be responsible for creating your own product system: Low Tech, High Touch and High Tech, High Touch products. Products do not live in a vacuum. They are used by a range of people, serve specific tasks, live in particular places, and require prior and/or learned knowledge to effectively use them. You are required to develop a device that has evolved through the years from a simple mechanical device to a more advanced device—but both accomplish the same task. For example, the simplest can opener is a fixed, hooked blade that requires several motions

Fig. 6.5 Securing and controlling nautical vessels, Roe, B. (2004). Georgia Institute of Technology

and power that comes from the user. However, electric can openers accomplish the same task but require energy, motors, and gears. Both devices were created for the same purpose but with very different methods, means, and materials.

There are endless types of products and categories that should be of interest to you. Alternately, there might be a low-tech product that has not evolved into a high-tech product. Think of problems other people have, common products that have changed over time, specific or specialized tasks, or products/problems that need addressing. You will uncover these through the process of your research and discovery stages.

The two redesigned or newly designed products should consider the type of interactions between the user, goals, and task/s. The chosen form and relationship of forms should relate to the function and past cognitive knowledge of the intended users. A sense of common brand should be apparent from both designs—just as a corporate structure, you will be responsible for creating a pair of products which are part of a family and system.

Having conducted this type of open-ended project challenge over 20 years, with all sorts of product categories, requires a depth of experience and knowledge from the instructors. This is due to the range of product categories which can span from soft goods to white box appliances, and use everything from simple mechanisms to digital. Example innovations of this project brief include a range of solutions from mapping and navigation, securing a nautical vessel, hydration while off-road cycling, photography and video equipment, kitchen tools, and household tools. The following Figs. 6.5, 6.6, and 6.7 are traditional industrial design examples of hard good products satisfying the original open-ended design brief.

Adjustment to the project brief's word terminology can include soft tech or soft materials. Or it might emphasize software if the intent is working in the digital domain. The design brief moniker might be recast as Soft goods, Software and Hard goods, Software. But essentially the foible between both products is the similarities and differences between the interactions as the goal is being completed, the cognitive differences of the person and the object interactions, and choices being made by the student.

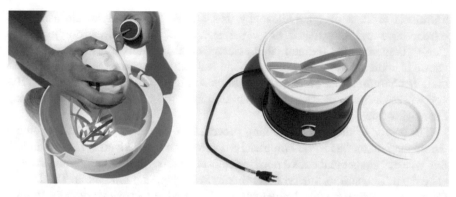

Fig. 6.6 Mixing and introducing air into food ingredients, Ellis Huff, H. (2004). Georgia Institute of Technology

Fig. 6.7 Controlling and stabilizing photo and video capturing, Newby, J. (2013). Carnegie Mellon University

Short of enrolling non-studio-trained employees into a semester-long design school, there are alternate ways of infusing these design challenges into a company. Several companies and institutions have intensive design workshops which can be extremely useful in introducing and orienting non-designers to the potential of the design processes. Rather than describing this as design thinking, rather the processes and methods should be regarded in this manner: design as thinking, design as learning, design as doing, and design as creating, and design as a means to better decision-making. Rather than imagined it as a thing, design thinking should be considered as an activity and part of the culture. To build this culture requires champions who have first-hand experiences. These growth mindset attitudes acknowledge that design thinking does not portray itself as a silver bullet or panacea to all problems. Rather it is a means for iterative methods to understand current states and developing plans for a future state. The human-centered design process, from a traditional product design philosophy, has always been humble, altruistic, and introspective and reflective. It has never been about ego or advertising grandiose claims. Contemporary literature that asserts that design thinking is the secret innovation ingredient is typically business and technology sectors. Designers are honored and humored that certain sectors think so highly of the design thinking. But you will see as many articles and literature from studio-trained designers tempering the magic of design thinking.

The power of the design thinking process is a hands-on, experiential activity where the person has to be completely involved and participating. Only through this type of engagement does a newly inculcated non-studio-trained student experience the full benefit of this type of thinking, working, and imagining. The Design Center at Carnegie Mellon University has delivered numerous workshops to non-designer organizations. First, established by Head of the School of Design, Terry Irwin, the Design Center was imagined as both an entity serving internal university constituents and external facing organizations and businesses. The human-centered design workshops are run by a range of instructors who individually have more than 10 to 20+ years of professional and higher education experiences. The clients range from the corporate to military and other educational institutions. More specifically, the executive design education team served business schools, healthcare organizations, consumer electronics companies, and branches of the military establishment. The term design thinking is an attractive method for changing internal approaches to new and old challenges. And the majority of the participants want to understand the nuances and details of the design process. These workshops span from 1 day to a 5-day series that are a combination of lectures that oversimplify traditional the design discipline and hands-on exercises that are situated in real-world problems. Some clients do not ask for detailed customization to match their industry. Rather they want to understand and experience the process as a designer would approach a situation in order to frame the issues or ascertain the problem. Similar to the studio project brief, the design charge is open-ended. We do not specify the problem nor dictate the parameters. We present the environment where the participants will observe work and everyday jobs and interactions. And through newly learned ways of observing, seeing, and recording, teams return with raw information. Teams then visualize the information to decipher the hidden opportunities through staged exercises that structure the initial unrelated research to find patterns and linkages. Debriefing post design thinking workshop, clients immediately understand the boots on the ground and hands-on necessity to develop a nascent sense for the double diamond process, human empathy, and use of visual mediums to communicate and collaborate.

Jon Kolko, Austin Center for Design founder wrote a book titled "Exposing the Magic of Design (2015)." He is a design practitioner and educator bringing both d/ Design to the corporate world and reimagining academia in his own way. Many people would quickly scour the book searching for the summary statement or the five to seven bullet points for the secrets to Design. Readers are quickly disappointed that it was not this easy. Each chapter's content and methods would build upon and be applied to the human-centered principles and philosophies proselytize in this text and many others before. The moral of every good practicing design text is that there are no short-cuts. The prior chapter case studies hopefully share just a glimpse into a team's highly contextual and generative activities and output.

Taxonomies of Problem Setting and Problem Sets

Delving into the sheer complexity of delivering an engineering curriculum through project-based learning is Dym, Agogino, Eris, Frey, and Le (2005) Engineering Design Thinking, Teaching, and Learning article. As one of the more comprehensive texts that hold an immense depth of experiences and prior literature, this piece highlights many of this book's tenor and intent. Dym et al. acknowledge that several concepts are at play when educating undergraduate engineering students. Design thinking, designing systems, design decisions, team environments, and the languages of engineering design are the main sections that lead toward best practices in project-based learning curriculum. Both at the introduction and end of the paper, a significant discussion is devoted to explaining and framing where design thinking plays a critical role in design engineering. One of the paragraphs of the first section titled, Design Thinking as Divergent-Convergent Questioning states:

Teaching divergent inquiry in design thinking is neither recognized clearly nor performed well in engineering curricula. For example, it is not acceptable for a student to respond to a final exam question in an engineering science course by providing multiple possible concepts that do not have truth value. Indeed, students are expected to engage in a convergent process by formulating a set of deep reasoning questions and working to the (unique) answer. Students' ability to converge is being positively assessed when partial credit is given for the "thought process," even if the answer is wrong. In this context, engineering curricula may be characterized as follows:

One of the main strengths of engineering curricula is their perceived effectiveness in conveying Aristotle's epistemological, convergent inquiry process. It promotes the ability to reason about knowledge associated with mathematics and sciences, which is construed as the engineering science or reductionist model.

Divergent inquiry takes place in the concept domain, where concepts or answers themselves do not have truth value, that is, they are not necessarily verifiable. This is the design or synthesis model. It often seems to conflict with the principles and values that are at the core of the predominantly deterministic, engineering science approach. The foregoing discussion raises the following question: Can the now more-formal identification of both divergent thinking and design as an iterative divergent–convergent process be used to develop better pedagogical approaches to both engineering design and engineering analysis?

This and works of several other authors (Vincenti, 1990; Adams et al., 2004; Cagan & Vogel, 2001) speak directly to the necessity of addressing skills for interdisciplinary collaboration, open-ended, project-based problems, and earlier in the curriculum. Dym cites Tooley and Kevin (1999) as recognizing established ABET accreditation criteria such as abilities to design a system, component, or process, work on multidisciplinary teams, social and ethical responsibilities, and global and social impact. These learning activities are seen not as just technical requirements but social activities that project-based cornerstone and capstone courses can and need to address. Measuring whether these non-technical skills and design thinking has value is impossible. However, Dym lists one of the positive aspects of introducing project-based learning as an early cornerstone course increases engineering student retention and inherent motivation in the major. Subsequently, it enhances performance in the capstone or later upper level or alternate courses that require open-ended, project-based challenges. Both anecdotal and published evidence from several prior papers support these positive effects (Pavelich & Moore, 1996; Adams, Turns, & Atman, 2003; Olds & Miller, 2004; Richardson & Dantzler, 2002; Hoit & Ohland, 1998; Willson, Monogue, & Malave, 1995).

Donald Schön, MIT professor wrote several books on the theory and practice of educating the professional practitioner. One his texts, "Educating the Reflective Practitioner: Toward a New Design for Teaching and Learning in the Professions" introduces the reader to the crisis of confidence in professional knowledge. Without the benefit of the double diamond model when he wrote his text, his process can be viewed as a dynamic repertoire of divergence and convergence. Schön went further to develop a framework that shows the range of cognitive shifts between design process stages (Schön, 1990).

1. problem situation – exploration, bounding and characterization of the problem in context
2. problem setting – determination of the programmatic goals, strategies, resources, etc., for project definition/go-ahead
3. problem framing – delineation of the essential technical issues and implications to be addressed, along with reservations and success criteria
4. problem specification – particular requirements that the design effort is committed to satisfy and verify.

This list could be interpreted as a student's design engineering's guiding template for a tackling project-based, open-ended challenges. Schön recognized that designers need to constantly balance between the rational while being able to embody that cognition into action and output. This, in turn, allows for another stage of reflection that begets new and deeper questions about the current solution state and whether it satisfies the original problem situation, setting, and framing. Some engineering academics assign value. to reflective practices by developing taxonomies of questions. Eris, 2003 paper on Asking Generative Design Questions: A Fundamental Mechanism in Design Thinking poses that, question asking is a fundamental cognitive mechanism in design thinking, and can be treated as a process. And the nature of questions engineers have while designing, is indicative whether deeper and more complex cognition is being undertaken (Eastman, Michael McCracken, & Newstetter, 2001). An in-depth investigation has been conducted by Robin S. Adams, Jennifer Turns, and Cynthia J. Atman in the seemingly obtuse area of design learning in engineering. In their paper, "What could design learning look like?" Adams, Turns, and Atman (2003) categorize to make sense of the process and students movement through their higher education experiences. They describe the Four Windows into Learning that is based on empirical studies of engineering student behaviors as:

A Design Process Window which illustrates within-subjects data of engineering student design processes

An Adaptive Expertise Window which illustrates multiple pathways to expertise

A Systems Window which illustrates how human learning can be represented as a complex dynamic system

A Writing as Design Window which illustrates learning as discourse

They further the depth of description by understanding the implications for developing design expertise continuum in the additional four areas:

Dimensions: What are the dimensions (or road markers) we can use to characterize the learning of engineering design and the acquisition of design expertise?

Shapes: How can we describe the shape of learning trajectories in the continuum?

Additional Insights: What additional insights are important in order to develop a design expertise continuum?

Implications: What are the implications for additional research?

The value to the product design field is that there exist very few if no studies with these type of protocols that allow for a basis of research and view into the complex subject of the design process. Identification of distinct areas, recognition of patterns, and enabling insights and further study for effective learning and education is essential if the world perceives design, design thinking, and the design process as a primary element to educating resourceful, collaborative, agile, and innovative designers and engineers. There exists many other sources and authors working in this area, however these sources act as prime examples of a necessary reductionist method to understand a complex process, system, and effect. There are some ambitious cognitive studies such as Cagan et al., who are attempting to see into the mind and brain of engineers through MRI scans and analysis (Cagan et al., 2013). Until that happens with a common consensus of protocols, results, and interpretations, we will continue to build through multi-method approaches, versions of the what and how a universal definition or even representation of the design process alludes us.

The value of an engineering education is not being challenged here. It is the need for adjustments and changes to engage students early and ongoing ways that amplify their current depth of knowledge. Though one absence between these engineering introspection analysis is mentioned in Chap. 1 design point of view, empathy. The primary constituent for a designer or engineer is the human. Though a civil engineer designing a bridge or dam may not directly impact an individual user on a daily basis, the scale in which their final designed artifact will most definitely impact many if not thousands of people in indirect ways. The gravity of this responsibility begets a process that does include the human or in this case, the many humans and the resultant social system and subsequent systems that are affected. Traffic patterns, new transportation infrastructures, sources of the construction materials, shifts in natural environments and the habit of animal species will invariably be affected. The range and depth of stakeholders need to be understood at the outset and through the design process. Thus teaching the crux of people-centered design and that planned intent through the process and product is essential.

Though Dym and Tooley earlier allude to in the social aspect as consideration for the human condition as paramount. So not all is lost in translation and the need to evaluate and teach future generations. The practice of design in any field requires input, analysis, and output to appropriately fit the context of that particular situation. Teaching this in theory and practice to young designers will prepare them for the complexities of the artificial world we are building for ourselves. Because in the end, every artifact we produce is affecting the human condition. The scale and amount of impact in which these weigh, are up to us to determine.

Case Study: Applying the Design Process to Imagine New Health Care Futures

I had the unique fortune of advising a Masters of Design graduate student who focused on improving the quality of life for home dialysis patients. Her undergraduate degree was in engineering. She had worked professionally in the biomedical field for several years before entering the Masters of Design graduate degree program. One of the main driving factors to pursue this degree and thesis was her personal observations and revelation that the machine itself was so stressful on the human psyche. One particular patient on home dialysis opted out of the process, essentially choosing to end his life because the process and life around him had so significantly changed his will to live. One has to pause to digest this mortal fact. But as any good problem solver, whether an engineer, designer, or doctor, the re-examination of what is causing this dire effect needs to be thoroughly investigated. A dialysis machine replicates the primary function of the kidneys to filter waste from the body and balances certain chemicals and hormones essential to the body. There are generally two types of options for patients requiring dialysis. Visit a clinic that is operated by technicians and nurses, or perform the same function at home. The graduate student focused on the in-home peritoneal dialysis treatment. Compared to the clinic, the perception was the patient has some control over when to conduct treatments. Additionally, there was a greater potential for possible change or impact in a non-clinical environment.

From a technical standpoint, safely and repeatedly filtering waste from a person's body is an incredible process of trial and discovery. The first iteration of hemodialysis was invented in 1943 by Dr. Willem Kolff. These systems clean a patient's blood through the fistula, graft, or catheter access. An external machine filters and returns properly pH-balanced blood back to the patient. This process requires travel to a health facility and professional caregivers present for the approximately 4-h process, three times a week. In 1968 Henry Tenckhoff created a new alternative at-home option called peritoneal dialysis. This process allows patients to self-administer care by using a permanent catheter placed in the stomach. Instead of filtering and cleaning the blood on the outside of the body, peritoneal dialysis uses a dialysate fluid to exchange waste product from within the body. Waste products

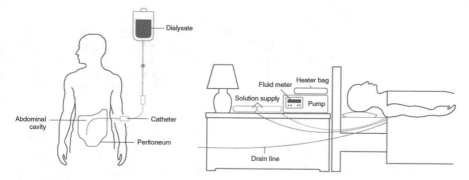

Fig. 6.8 Diagram of peritoneal dialysis

through capillary action move across semipermeable membranes to the dialysate located in the patient's peritoneal space. Two options for peritoneal dialysis exist: continuous ambulatory peritoneal dialysis is without a machine, and automated peritoneal dialysis is with a machine.

Continuous ambulatory peritoneal dialysis consists of hanging a dialysate bag and having gravity push the fluid through the peritoneal space and out into a drainage bag. However, the process is involved as it requires up to four exchanges of dialysate fluid. The exchange itself can take 30–40 min. However, the dwell time or time in which the dialysate fluid stays in the patient's belly is 4–6 h. At sleep time, the dialysate fluid stays in the patient's belly.

The machine automated peritoneal dialysis is typically employed during a patient's sleep period. This system cycles or fills and empties the dialysate fluid from the belly three to five times during the nighttime. And once the patient is awake, they start with a daylong dwell time. After effects of these treatments can include nausea, headaches, cramping, and feel fatigued after the treatment.

Patients submit to a laborious and life-altering regimen that requires being tethered and beholden to external mechanisms. This can be seen as an inconvenience if there were no other alternative. Figure 6.8 shows the elements of the system. However, to experience the home dialysis process over an unknown length of time (waiting for a kidney transplant donor match) is hard to personally conceive. If a designer cannot experience it first-hand, then they attempt to understand and empathize with and through them. This student devoted significant time and processes to discover the mundane to the nuanced when considering the people she was trying to help. Behavioral and generative research methods such as observations, first-person interviews, and participatory methods were utilized to form a more complete picture of the peritoneal home dialysis regiment process. Figure 6.9 shows a single day's supply requirements. Figure 6.10 shows the touchpoints and interactions that a patient has to experience for each treatment.

The logistics of the life-saving process has become an animal onto itself. The enormity of managing and stocking the supplies becomes a part-time job. Add to this the actual time in which a patient has to undergo their own treatment, and one

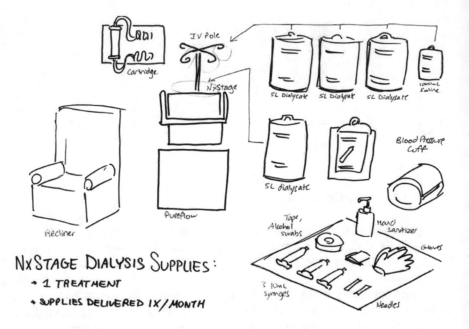

Fig. 6.9 Supplies required for single day home peritoneal dialysis treatment. Impellicceiri (2015). Carnegie Mellon University

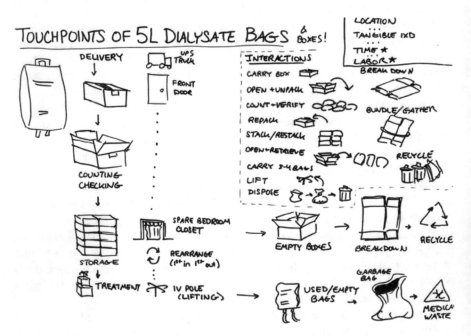

Fig. 6.10 Touchpoints and interactions required for each self-treatment, Impellicceiri (2015). Carnegie Mellon University

can quickly discern the inequities. The patient has now become a slave to the process. And without feedback or a method for predicting a kidney donor match, hope and the will to live does become a question for some.

Changing one's perspective is an essential part of the design process. This graduate student did propose new physical and digital product system to provide more direct feedback to the patient. But the other proposal was more about changing and positively affecting the quality of life. And this was about the management and service of all the collateral supplies, dialysis, waste by-product, and schedules. Even though an engineer has all the technology and knowledge (mechanical, chemical, electrical, fluid dynamics, etc.) necessary to save a human for prolonged periods of time, there exists the very human aspect of living. This case highlights that good technical engineering is only part of the equation to delivering sustainable and usable solutions.

References

Adams, R. S., Turns, J., & Atman, C. J. (2003). Educating effective engineering designers: The role of reflective practice. *Design Studies, 24*(3), 275–294.

Adams, R. S., Turns, J., Martin, J., Newman, J., & Atman, C. (2004). An analysis on the state of engineering education. unpublished manuscript intended for submission to the *Journal of Engineering Education*.

Cagan, J., & Vogel, C. (2001). *Creating breakthrough products: Innovation from product planning to program approval*. New Jersey: FT Press.

Cagan, J., Dinar, M., Shah, J. J., Leifer, L., Linsey, J., Smith, S. M., & Vargas-Hernandez, N. (2013). Empirical studies of design thinking: Past, present, future. In *Proceedings of the ASME Design Engineering Technical Conference* (Vol. 5). New York, NY. [V005T06A020].: American Society of Mechanical Engineers. https://doi.org/10.1115/DETC2013-13302.

Dym, C. L., Agogino, A. M., Eris, O., Frey, D. D., & Le, L. J. (2005). Engineering design thinking, teaching, and learning. *Journal of Engineering Education, 94*, 103–120.

Eastman, C. M., Michael McCracken, W., & Newstetter, W. C. (2001). *Design knowing and learning: cognition in design education*. Amsterdam: Elsevier Science B.V.

Eris, O. (2003). Asking generative design questions: a fundamental cognitive mechanism in design thinking. International Conference on Engineering Design.

Georgia Institute of Technology. (n.d.). *School of Engineering website*. Retrieved February, 2018, from http://www.me.gatech.edu/about/history.

Hoit, M., & Ohland, M. (1998). The impact of a discipline-based introduction to engineering course on improving retention. *Journal of Engineering Education, 87*(1), 79–85.

Kolko, J. (2015). *Exposing the magic of design: A practitioner's guide to the methods and theory of synthesis, Human technology interaction series*. New York, NY: Oxford University Press.

Olds, B. M., & Miller, R. L. (2004). The effect of a first-year integrated engineering curriculum on graduation rates and student satisfaction: A longitudinal study. *Journal of Engineering Education, 93*(1), 23–35.

Pavelich, M. J., & Moore, W. S. (1996). Measuring the effect of experiential education using the Perry model. *Journal of Engineering Education, 85*(2), 287–292.

Richardson, J., & Dantzler, J. (2002). Effect of a freshman engineering program on retention and academic performance. *Proceedings, Frontiers in Education Conference, Institute of Electrical and Electronic Engineers*, S2C-16-S2C-22.

Schön, D. (1990). *Educating the reflective practitioner: Toward a new design for teaching and learning in the professions* (1st ed.). San Francisco, CA: Jossey-Bass.

Tooley, M.S., & Kevin, D. (1999). Using a capstone design course to facilitate ABET 2000 program outcomes. *Proceedings, ASEE Conference & Exhibition*, Session 1625.

University of Michigan. (n.d.). *Mechanical Engineering website*. Retrieved February, 2018, from https://me.engin.umich.edu/about/history.

Vincenti, W. G. (1990). *What engineers know and how they know it: Analytical studies from aeronautical history*. Baltimore, MD: Johns Hopkins University Press.

Willson, V., Monogue, T., & Malave, C. (1995). First year comparative evaluation of the Texas A & M Freshman Integrated Engineering Program. *Proceedings, Frontiers in Education Conference Institute of Electrical and Electronic Engineers*.

Index

© Springer Nature Switzerland AG 2019
W. C. Chung, *The Praxis of Product Design in Collaboration with Engineering*,
https://doi.org/10.1007/978-3-319-95501-8